Cartilaginous Conquerors: The Enduring Legacy of Chondrichthyes

Kinky

Table of Contents

Chapter One

General Introduction

The evolution and diversity of Chondrichthyes

Chondrichthyes is an evolutionary conserved and successful class of jawed vertebrates, representing one of the two groupings of extant fish (Carroll, 1988; Simpfendorfer and Heupel, 2012), forming a monophyletic sister-taxon to Osteichthyes. Due to their possession of jaws, Chondrichthyes and Osteichthyes are termed the crown Gnathostomata (Motta *et al.*, 1997; Meyer and Zardoya, 2003; Carrier *et al.*, 2012; Boisvert *et al.*, 2019). The class Chondrichthyes includes sharks, skates, rays and chimaeras (Thompson and Springer, 1965; Heinicke *et al.*, 2009; Carrier *et al.*, 2012; Ebert and Stehman, 2013; Walker, 2020).

Chondrichthyans made their first appearance during the late Silurian period, over 400 million years ago, and are presumed to have derived from the Acanthodians (Compagno, 1990b, 1999; Brett and Walker, 2002; Last and Stevens, 2009; Grogan *et al.*, 2012; Hastings *et al.*, 2015). The earliest record of chondrichthyans is "chondrichthyan-like scales" (Carrier *et al.*, 2012). This finding suggests that early species may not have been well mineralised, which results in the lack of fossils found today (Carrier *et al.*, 2012). While other faunal groups were dying out, chondrichthyans radiated during the Permian-Triassic and Cretaceous-Triassic periods of mass extinction and have shown spikes in their diversity, increasing and decreasing over time (Carroll, 1988; Fowler *et al.*, 2005; Carrier *et al.*, 2012; Hastings *et al.*, 2015). These fishes have occupied and survived in diverse ecosystems, and have shown remarkable evolutionary plasticity and resilience with time (Fowler *et al.*, 2005; Carrier *et al.*, 2012).

Presently, Chondrichthyes is comprised of 14 Orders, over 60 Families, 188 Genera, and about 1200 species, with many species still being described (Compagno, 1990b; a; Last and Stevens, 2009; Grogan *et al.*, 2012; Hastings *et al.*, 2015). Even though chondrichthyan fishes comprise only 3% of the total fish diversity today, this class includes more than 10% of the known fish Orders, highlighting remarkable differences in morphology among relatively few species (Last and Stevens, 2009; Hastings *et al.*, 2015). This class is split into two divergent evolutionary lines (sub-classes), namely the Holocephali (complete head - chimaeras; elephant fishes and ratfishes) and Elasmobranchii (strap gills - sharks, skates and rays) (Lund and Grogan, 1997): Elasmobranchii comprises 96% of the overall diversity (Compagno, 1999; Klimley, 2013; Hastings *et al.*, 2015).

Unique biology and general ecology of Chondrichthyes

Chondrichthyan fishes are distinguished from Osteichthyes by possessing a skeleton made of cartilage instead of bone, and a neurocranium without sutures (Carrier *et al.*, 2012; Hastings *et al.*, 2015). The cartilaginous skeleton gives the group its alternative name and they are more commonly referred to as cartilaginous fish. They have evolved well-developed electro-receptive senses, with numerous pores of the ampullae of Lorenzini. These receptive senses function in detecting weak electric fields emitted by prey, thus allowing for the detection of prey in the absence of olfactory and visual cues (Klimley, 2013). The ampullae of Lorenzini are especially well-pronounced around their mouth region (Klimley, 2013; Hastings *et al.*, 2015).

All species of cartilaginous fish undergo internal fertilization (Ebert and Stehman, 2013; Klimley, 2013). Depending on the species, females may be oviparous and release protective keratinized egg-cases in which their embryos develop; while others may be ovoviviparous - with embryos developing within the body of the females (Hastings *et al.*, 2015). Within ovoviviparous species, some embryos develop exclusively from nutrition supplied in the egg (yolk-viviparity); others develop from inter-uterine cannibalism; while the mothers of some species provide nutrition to developing embryos, either by placental viviparity, oophagy or lipid histotrophy (Hamlett, 2005; Musick, 2011). Musick and Ellis, (2005) concluded that the primitive state in chondrichthyans was yolk-sac viviparity, from which all other forms of ovo-viviparity had evolved. The males are modified for internal fertilization, as they possess a pelvic-fin clasper that aids in mating. Its sole function is to transfer sperm during mating (Klimley, 2013). Even though some species of chondrichthyan fish may have well-defined reproductive seasons, others do not and it is often difficult to determine the reproductive cycles for many species (Klimley, 2013). Importantly, the reproductive output of chondrichthyan fish tends to be limited, with small litter sizes being the general condition for most species (Klimley, 2013; Hastings *et al.*, 2015). As a consequence, they are regarded as having *k*-selected life-history strategies (Klimley, 2013), characterised by slow growth, late age at maturity, low fecundity, long gestation periods and long lives (Compagno, 1990b; Klimley, 2013). This suite of life-history characteristics allows for a low intrinsic rate of population increase, and consequently, a low reproductive potential, making them vulnerable to overexploitation (Compagno, 1990b; a; Stevens *et al.*, 2000; Baum, 2003; Cavanagh and Gibson, 2007; Ferretti *et al.*, 2010; Worm *et al.*, 2013).

Cartilaginous fish display a diverse range in size, with some fully-grown at ≤ 20 cm (e.g. *Squaliolus aliae*) and others only fully-grown at sizes ≥ 500 cm (e.g. *Rhincodon typus*; Colman, 1997; Kyne and Heupel, 2015; Pierce and Norman, 2016). Most species are predominantly found in marine environments (Klimley, 2013) and may occur either along continental shelves or continental slopes, with few species known to even occur along insular shelves. Even though some species may occur in inshore estuarine and freshwater habitats, only a few species have ranges that are restricted to such environments (Klimley, 2013). Estuarine and freshwater habitats are generally known to serve as birthing grounds and nurseries (Carrier *et al.*, 2012; Klimley, 2013). Chondrichthyan fish are known to occur at all depths in the ocean, and even though they are highly mobile organisms, many have restricted distributions with only a few taking part in large-scale migrations (Cailliet *et al.*, 2005; Cavanagh and Kyne, 2006).

Chondrichthyans are entirely predacious and are well-known carnivorous hunters (Klimley, 2013). The teeth of sharks, skates, and rays are derived from placoid scales, which form replicating rows that are serially replaced. Chimaeras, on the other hand, have three pairs of permanent, grinding, non-mineralised tooth plates (Helfman *et al.*, 2009; Lisney, 2010; Hastings *et al.*, 2015). Although some species may feed by filter-feeding (e.g. Rhincodon typus), most are regarded as predators and many occupy top or near-top roles in the food web (Compagno, 1990a; Stevens *et al.*, 2000; Klimley, 2013). This top-predatory position is apparent, as they play key roles in predator-prey relations by affecting the population size of the prey species, thereby greatly influencing ecosystem structure and function (Stevens *et al.*, 2000; Klimley, 2013).

The Elasmobranchs

Individuals belonging to the sub-class Elasmobranchii are characterised by several features including five to seven separate gill openings, a dorsal fin that is entirely rigid, spiracles, blind sac nostrils, a body that is covered in placoid scales with either amphistylic (upper jaw not fused to the cranium) or hyostylic jaws (upper jaw fused to the cranium; Hastings *et al.*,

2015). In addition to this, males within this group lack cephalic claspers (Klimley, 2013). Elasmobranchii is comprised of 13 orders namely, Heterodontiformes (horn sharks), Orectolobiformes (nurse sharks), Carcharhiniformes (ground sharks), Lamniformes (mackerel sharks), Squaliformes (dogfish sharks), Hexanchiformes (cow sharks), Squantiniformes (angel sharks), Pristiophoriformes (saw sharks), Echinorhiniformes (bramble and prickly sharks), Rhinopristiformes (sawfishes), Rajiformes (skates), Torpediniformes (electric rays) and Myliobatiformes (stingrays; Klimley, 2013; Hastings *et al.*, 2015; Fricke *et al.*, 2020).

While most collectively refer to the group as elasmobranchs, phylogenetic analyses have shown further split in this relationship, with the skates and rays comprising the group, Batoidea and the shark-like elasmobranchs comprising the sister-group, Selachii (Naylor *et al.*, 2005; Aschliman *et al.*, 2012). Interestingly, the numbers of extant elasmobranch species are almost evenly divided between Batoidea (54% of the total chondrichthyan diversity) and Selachii (42%) (Compagno, 1999; Hastings *et al.*, 2015). Batoidea are characterized by having dorso-ventrally flattened heads and bodies; enlarged pectoral fins adjacent to the head, mouth, and gills; gill slits opening ventrally and the eyes and spiracles positioned on the dorsal region of the head (Hastings *et al.*, 2015). By contrast, Selachii are characterized by having pectoral fins separate from the head and gills; gill-openings on the lateral side of their head, with eyes and spiracles positioned in the anterior region of the head (Nelson *et al.*, 2016; Hastings *et al.*, 2015).

The pronounced *k*-life-history characteristics of elasmobranchs highlight their intrinsic vulnerability (Compagno, 1990b; Klimley, 2013). Numerous studies have already indicated worldwide declines in elasmobranch populations as a result of direct and indirect anthropogenic activities (Stevens *et al.*, 2000; Dulvy and Forrest, 2010; Hutchings *et al.*, 2010; Barausse *et al.*, 2014). Unsustainable fishing practices and the consequent bycatch activities have been leading causes of widespread depletion of elasmobranch populations (Stevens *et al.*, 2000; Frisk *et al.*, 2001; Carrier *et al.*, 2004; Frisk, 2010). In addition to this, the continued expansion of fishing into the deep-sea has raised concerns about the ability of deep-sea chondrichthyans to sustain fishing pressure (Morato *et al.*, 2006). Of the global fauna of about 1200 described species of chondrichthyans, approximately 530 are deep-sea species (Hastings *et al.*, 2015), with most deep-sea chondrichthyans yet to be studied (Frisk, 2010). Therefore, urgent studies on the distribution, life history, population dynamics and feeding habits of these species are required, as those currently available mainly deal with species of commercial importance.

In most regions of the world, squaloid (dogfish) sharks are of no commercial importance but are readily removed in substantial quantities as bycatch (Compagno *et al.*, 2005; Petersen *et al.*, 2008; Ebert, 2013; a; Ebert and Stehman, 2013; Oliver *et al.*, 2015; Currie *et al.*, 2020). In the few regions where dogfish are of commercial importance, these sharks are targeted for their meat, liver, oil, fins and sometimes leather (Compagno *et al.*, 2005; Ebert, 2013). Around the South African coastline, studies on commercially unimportant sharks are lacking, even although they do make a substantial contribution to the bycatch (Petersen *et al.*, 2008; Reed *et al.*, 2017; Currie *et al.*, 2020).

Dogfish taxonomy, distribution, and biology

The family Squalidae consists of two genera, *Cirrhigaleus* and *Squalus,* which collectively comprises 36 species (Ebert and Stehman, 2013; Froese and Pauly, 2019). Over the years, there has been a tendency to group morphologically similar, but geographically separated *Squalus* populations in global "species complex sub-groups" (Leslie RW, *pers comm.,* 2015, DFFE), namely, *S. acanthias, S. megalops* and *S. mitsukurii* (Ebert *et al.*, 2010; White and Iglésias, 2011; White *et al.*, 2013; Viana *et al.*, 2017). Recently, however, the genus *Squalus* has received a lot of attention, as many new species have been described and older taxa have been resurrected (Last *et al.*, 2007; Ebert *et al.*, 2010; Figueirêdo, 2011; Veríssimo *et al.*, 2011; White and Iglésias, 2011; White *et al.*, 2013; Viana and de Carvalho, 2016; Viana *et al.*, 2017). While some studies have made use of morphology as the only aid in identifying species, others have also employed genetic techniques (Viana and de Carvalho, 2016; Veríssimo *et al.*, 2017; Viana *et al.*, 2017). Unfortunately, the taxonomy of *Squalus* remains unresolved in most regions of the world, and there have been recent spikes in taxonomic studies of *Squalus* species in southern African waters (Viana and de Carvalho, 2016; Veríssimo *et al.*, 2017; Viana *et al.*, 2017).

Squalus sharks have short-nosed cylindrical "shark-bodies". The presence of an un-grooved spine in front of their first and second dorsal fins, as well as the absence of an anal fin, are the key features used to distinguish this family in the field (Ebert and Stehman, 2013). Other diagnostic features used to further distinguish members within this genus include a broad and a flat head; five gills of equal size; short transverse mouth; powerful jaws with sharp cutting teeth in their upper (21 – 30) and lower (21 – 27) jaw; the presence of an upper precaudal pit and the absence of a sub-terminal notch on the caudal fins (van der Elst, 1993; Heemstra and Heemstra, 2004; Ebert and Stehman, 2013).

Dogfish colours range from grey to brown and bronze, with some having prominent markings on their body (Ebert and Stehmann, 2013). Like all other Chondrichthyes, reproduction in species of *Squalus* requires internal fertilization, with their mode of reproduction being ovoviviparity. Females produce between one and 32 pups per litter after a gestation period of between 18 - 24 months (van der Elst, 1993; Heemstra and Heemstra, 2004; Ebert, 2013). The size at which maturity is reached ranges from between 60 cm and 120 cm in length (Ebert, 2013). Females mature at larger sizes than the males, and all squaloid species are slow-growing, with some only maturing after 30 years: some species can live up to 100 years (Ebert, 2013; Ebert and Stehmann, 2013). Dogfish are limited to the tropical and temperate regions of the world's oceans and are known to live around islands, continents, submarine peaks and ridges and are also known to live in close association with the sea-floor (benthic or demersal; Compagno, 1984; Compagno *et al.*, 2005; Ebert, 2013; Ebert and Stehman, 2013). Most dogfish are regarded as deep-sea species and occur at depths greater than 200 m, though a few can be found at > 600 m (Compagno *et al.*, 2005). A few species are also known to occur in the water column (pelagic), with some even shown to inhabit shallow water embayments along sandy beaches and rocky reefs. Some species of *Squalus* are social and form large "packs", while others occur in smaller groupings, which are generally reported to segregate by size and sex (Heemstra and Heemstra, 2004; Compagno *et al.*, 2005; Ebert, 2013; Ebert and Stehmann, 2013). Other species remain solitary (Heemstra and Heemstra, 2004; Compagno *et al.*, 2005; Ebert, 2013; Ebert and Stehmann, 2013; Hastings *et al.*, 2015).

The diet of squaloids primarily consists of bony fish and invertebrates, though they may feed on other chondrichthyans and cetaceans (Ebert *et al.*, 1992). They are often regarded as generalist feeders (Ebert *et al.*, 1992; van der Elst, 1993; Heemstra and Heemstra, 2004; Compagno *et al.*, 2005; Ebert, 2013). The dogfish that occur in packs may be feeding communally and cooperatively work together to attack prey (Ebert, 2013).

Like most sharks *Squalus* sharks have a low intrinsic rate of population increase, making them vulnerable to overfishing (Ebert *et al.*, 1992; Gelsleichter *et al.*, 1999; Stevens *et al.*, 2000; Simpfendorfer and Kyne, 2009; Ebert, 2013; Dulvy *et al.*, 2014). Unfortunately, squaloid sharks are one of the most common bycatch species, being taken in mixed fisheries(Jakobsdóttir, 2001; Hutchings and Lamberth, 2002; Petersen *et al.*, 2008; Attwood *et al.*, 2011; Ebert, 2013; Ebert and Stehman, 2013; Reed *et al.*, 2017; Currie *et al.*, 2020). Although also taken as bycatch in pelagic trawls, gill/seine nets, they are most commonly caught as a bycatch of demersal trawl fisheries (Leslie RW, *pers comm.,* 2015, DFFE; Currie *et al.*, 2020).

In South Africa, *Squalus acutipinnis* (formerly considered a junior synonym of *S. megalops*) (Regan 1908) and *S. bassi* (formerly known as *S. cf mitsukurii*, Jordan and Snyder 1903) are two of the most common demersal bycatch species (Petersen *et al.*, 2008). The taxonomy of *Squalus* around southern Africa is complicated and debates are ongoing (Viana and de Carvalho, 2016; Veríssimo *et al.*, 2017; Viana *et al.*, 2017, Leslie RW, *pers comm.,* 2019, DFFE). At the time this study was initiated, the taxonomy of species around South Africa was unresolved, and *S. acutipinnis* was regarded as *S. megalops*, and *S. bassi* was considered to be
S. mitsukurii.

The shortnose dogfish of southern Africa

Regan (1908) had described the population of shortnose dogfish off South Africa as a distinct species, *Squalus acutipinnis*: Bass *et al.* (1976) designated this species as a junior synonym of *S. megalops*. In 2012, a study by Naylor *et al.*, concluded that the species, *S. cf megalops* from South Africa is genetically distinct from *Squalus megalops* (Naylor *et al.*, 2012). Eshmeyer's Catalogue of Fishes (Eschmeyer, 2014) then accepted *S. acutipinnis* (Regan, 1908) as a valid taxon and that has been used in various recent publications (eg. Dippenaar and Molele, 2015). Since then, Eshmeyer has changed his mind and has once again listed *S. acutipinnis* as a junior synonym of *S. megalops*, leaving a lot of taxonomic confusion around this species. This led to two studies that aimed at reassessing the phylogeny of the genus *Squalus* off South Africa, which in turn has led to the resurrection of *S. acutipinnis* (Viana and de Carvalho, 2016; Verissimo *et al.*, 2017).

Squalus acutipinnis has a distribution range from Angola to South Africa, occurring on or near the seafloor on the inner/outer continental shelves and the upper slopes, to a maximum of about 500 m (Heemstra and Heemstra, 2004; Ebert, 2013). The juveniles are pelagic and occur across the outer continental shelf (Watson and Smale, 1998, 1999;

Heemstra and Heemstra, 2004; Ebert, 2013; Dippenaar and Molele, 2015). *Squalus acutipinnis* can be recognized by its brown coloured upper body with a white underside. It is further distinguished from other squaloids by the length of the spines on the dorsal fins and the comparative distances between the nostril and the snout, to the nostril and the labial furrow. The spine of the first dorsal fin is considerably shorter than the height of its associated dorsal fin, while the length of the spine in front of the second dorsal fin is approximately the same as the height of this fin (Heemstra and Heemstra, 2004; Compagno *et al.*, 2005; Ebert, 2013; Ebert and Stehman, 2013; Ebert and Mostarda, 2016). The distance between the snout and the nostril is greater than that between the nostril and the labial furrow (Heemstra and Heemstra, 2004; Compagno *et al.*, 2005; Ebert, 2013; Ebert and Mostarda 2013, Ebert and Stehman, 2013; Ebert and Mostarda, 2016).

Squalus acutipinnis may attain a maximum size of roughly 77 cm in total length (TL), although few individuals are longer than 65 cm (TL) (Watson and Smale, 1998, 1999; Ebert and Mostarda 2013 Leslie RW, *pers comm.*, 2019, DFFE). Females reach maturity between 49 cm and 55 cm, while males mature at a length of 40 cm (Watson and Smale, 1998, 1999; Heemstra and Heemstra, 2004; Compagno *et al.*, 2005; Ebert and Mostarda, 2013; Ebert and Mostarda, 2016). The species is ovoviviparous, with females birthing between two and six pups during winter (Heemstra and Heemstra, 2004). After a two-year gestation period, pups measure between 23 – 24 cm (Watson and Smale, 1998, 1999; Heemstra and Heemstra, 2004; Compagno *et al.*, 2005; Ebert and Mostarda, 2013; Ebert and Mostarda, 2016). *Squalus acutipinnis* grows very slowly, with females reaching a maximum age of about 32 years and males 29 years. These sharks aggregate in large and dense schools that segregate by size and sex (van der Elst, 1993; Watson and Smale, 1998, 1999; Heemstra and Heemstra, 2004; Compagno *et al.*, 2005; Ebert and Mostarda, 2013; Ebert and Stehmann, 2013; Ebert and Mostarda, 2016).

A preliminary study by Ebert *et al.* (1992) indicated that *Squalus acutipinnis* feed predominantly on bony fish, with some even feeding on cephalopods and to a lesser extent on crustaceans. *Squalus acutipinnis* is a regular by-catch, caught by shore-anglers but more commonly by bottom trawlers (Ebert and Stehmann, 2013). Although having no commercial status in South Africa, some *S. acutipinnis* are reportedly filleted and exported (Heemstra and Heemstra, 2004; Compagno *et al.*, 2005; Ebert, 2013; Ebert and Stehmann, 2013). *Squalus acutipinnis* is regarded as a meso-predator within these waters and up until very recently possibly a regional endemic (Viana and de Carvalho, 2016; Leslie RW, *pers comm.*, 2019,

DFFE).

The short-spine dogfish of southern Africa

The taxonomy of *Squalus bassi* can also be described as confusing. Initially, this species was called *S. bassi* in these waters, however, years later got reassessed and then got grouped into the *S. mitsukurii* species complex and referred to as *S. c.f. mitsukurii* in this region (Ebert *et al.*, 1992). Years later again, Viana *et al.* (2017) resurrected the name *S. bassi*, as these authors found substantial differences between *S. mitsukurii* and *S. bassi*.

The short-spined dogfish off southern Africa is distributed from Namibia to Mozambique. These sharks occur on continental and insular shelves, along upper slopes as well as around submarine ridges and seamounts, at a maximum depth of about 1000 m (Jordan and Snyder, 1903; Ebert, 2013, 2015; Ebert and Mostarda 2013; Ebert and Stehman, 2013). *Squalus bassi* is distinguished from other species in the field by its pearl grey to grey-brown body and white underside (Ebert and Mostarda, 2013). It is distinguished from the rest of the squaloid species by having a relatively short-spine in front of the second dorsal fin, by comparison to the height of the fin itself. It is further distinguished by the short distance between the nostril and the snout relative to the distance between the nostril and the labial the furrow (Ebert and Mostarda, 2013).

Short-spine dogfish attain a maximum size of 103 cm (TL) (Lucifora *et al.*, 1999; Oddone *et al.*, 2010; Ebert, 2013; Ebert and mostarda, 2013, 2016; Ebert and Stehman, 2013). Females reach maturity at a length of 72 cm, while males mature between 65 and 89 cm (Lucifora *et al.*, 1999; Heemstra and Heemstra, 2004; Compango *et al.*, 2005; Ebert and Stehmann, 2013). The mode of reproduction is ovoviviparity, with females birthing between four and nine pups during autumn (Ebert and Stehmann, 2013). After a two-year gestation period, pups are born measuring between 22 – 26 cm (van der Elst, 1993; Heemstra and Heemstra; 2004; Compagno *et al.*, 2005). These sharks may aggregate in large, dense sexually segregated schools. A preliminary study by Ebert *et al.* (1992) has indicated that short-spine dogfish feed predominantly on bony fish and cephalopods, with some even feeding on crustaceans and polychaetes; they are regarded as mesopredators. *Squalus bassi* is a regular bycatch species, caught throughout its distributional range (Heemstra and Heemstra, 2004; Cailliet *et al.*, 2005; Compagno *et al.*, 2005; Cavanagh and Gibson, 2007; Ebert, 2013; Ebert and Mostarda, 2013, 2016; Ebert and Stehman, 2013), and is considered a regional endemic (Viana *et al.*, 2017; Leslie RW, *pers comm.*, 2019, DFFE).

Study motivation and objective

One of the biggest problems facing shark research, is that focus tends to be given to either the bigger and more charismatic nearshore species, or those of commercial value (Cortés and Gruber, 1990; Ebert *et al.*, 1992; Platell *et al.*, 1998; Stevens *et al.*, 2000; Linke *et al.*, 2001; Bethea *et al.*, 2004; White *et al.*, 2004; Rinewalt, 2007; Papastamatiou, 2008; Acuña and Villarroel, 2010; Vaudo and Heithaus, 2011; Kyne *et al.*, 2011; Mulas *et al.*, 2011; Dicken *et al.*, 2017). Lesser attention is given to the uncharismatic, mesopredatory sharks, which are more likely than not regarded as "Data Deficient" by the International Union of the Conservation of Nature (IUCN) Red List (Cortés and Gruber, 1990; Ebert *et al.*, 1992; Stevens, 2000; Bethea *et al.*, 2004; Kyne *et al.*, 2011; Mulas *et al.*, 2011; Vaudo and Heithaus, 2011; Lopez *et al.*, 2012).

Yet mesopredatory sharks represent intermediate trophic-level predators that form the link between the upper and lower trophic levels (Ritchie and Johnson, 2009; Vaudo and Heithaus, 2011). Given this important ecological role, an understanding of their distribution, biology and ecology is important as it provides information on community dynamics and ecosystem functioning (Heithaus *et al.*, 2002; Braccini *et al.*, 2005; Acuña and Villarroel, 2010; Vaudo and Heithaus, 2011; Brown *et al.*, 2012; Lopez *et al.*, 2012; Simpfendorfer *et al.*, 2012).

Fishing has a major impact on sharks, affecting populations through habitat disturbance, removal and ultimate death (Belleggia *et al.*, 2012). As fish have become increasingly scarce in shallow and accessible waters, the fishing industry has begun moving offshore and has started fishing in deeper waters, where the mesopredatory species are predominantly found (Ferretti *et al.*, 2010; Schoener *et al.*, 2012). While squaloids make-up the second most significant component of the demersal fauna (in terms of trawl diversity), around southern Africa (Watson and Smale, 1999; Richardson *et al.*, 2000; Veríssimo *et al.*, 2011; Dunn *et al.*, 2013), our knowledge of their intricate population dynamics, distribution, and ecology is depauperate.

This study aims to examine the distribution and diet of *Squalus acutipinnis and S. bassi* along the West and South coasts of South Africa, in order to improve the knowledge base upon which management decisions about bycatch are based. Furthermore, because *S. acutipinnis* and *S. bassi* are described as being similar in morphology, habit and biology and are known to co-occur, overlapping in latitudinal and bathymetric distributions (Ebert and

Stehmann, 2013), it is hoped to shed light on the nature of interactions between them. This study will be the first of its kind on these dogfish in this region of the world's oceans. Each data chapter will have its own introduction, materials and methods, results and discussion sections and are governed by the following set of objectives:

Chapter Two: Distribution

1. Determine sex-related differences in the distribution of *Squalus acutipinnis* and *S. bassi* around South Africa
2. Analyse fishery-independent data to determine the geographic (South or West coast) and bathymetric (Depth) distribution and Size composition, in space of *Squalus acutipinnis* and *S. bassi* around the West and South Coast of South Africa
3. Examine catch data to determine whether these species intraspecifically and interspecifically differ in the use of their physical habitat
4. Quantify and compare species density around the West and South coast
5. Compare the distribution of *Squalus acutipinnis* and *S. bassi* to that of the trawling effort, to estimate the potential for trawl impact

Chapter Three: Diet by stomach content analyses

1. Determine sex-related variability in the diet of *Squalus acutipinnis* and *S. bassi* that occur in the West and South coast waters of South Africa
2. Determine intraspecific variability in the diet of *Squalus acutipinnis* and *S. bassi* that occur in the West and South coast waters of South Africa, by Coast, Depth and Size class
3. Determine the interspecific variability in the diet between *Squalus acutipinnis* and *S. bassi* that occur in the West and South coast waters of South Africa, by Coast, Depth and Size class
4. Examine diet data to determine if *Squalus acutipinnis* and *S. bassi* differ in the use of food resources, intraspecifically and interspecifically, in the waters of the West and South coast of South Africa

Chapter Four: Trophic ecology by stable isotope analyses

1. *Determine* sex-related differences in the isotopic signatures of *Squalus acutipinnis* and *S. basis* that occur in the waters of the West and South coast of South Africa

2. Determine intraspecific variability in isotopic signatures of $\delta^{15}N$ and $\delta^{13}C$, by Coast, Depth and Size classes for both *Squalus acutipinnis* and *S. bassi* that occur in the waters of the West and South coast of South Africa

3. Determine interspecific differences in isotopic signatures of $\delta^{15}N$ and $\delta^{13}C$, by Coast, Depth and Size classes, between *Squalus acutipinnis* and *S. bassi* that occur in the waters of the West and South coast of South Africa

4. Examine isotopic data to determine whether these species differ intraspecifically and interspecifically, in the use of their trophic habit

Chapter Two

The distribution and population structure of the dogfish, *Squalus acutipinnis* and *Squalus bassi* on the West and South coasts of South Africa

Introduction

The manner in which species are non-randomly distributed within their environments is a fascinating aspect of ecology and has long intrigued ecologists (Guisan and Thuiller, 2005). Multiple studies carried out within the marine environment have indicated significant spatial and temporal variation in the abundance and distribution of fish, within their respective communities (Diaz, 2014). Depending on their size, sharks are regarded as either apex predators (top) or meso/near-top predators (mid-trophic level) and have been shown to have strong regulatory functions in marine ecosystems (Heithaus *et al.*, 2002; Papastamatiou, 2008). Through their interactions, sharks may control the abundance of prey within their community, thus affecting the functioning and structure of their associated ecosystems (Ceccarelli *et al.*, 2014; Diaz, 2014).

Species distribution patterns, witnessed today, reflect multiple abiotic and biotic factors (Knip *et al.*, 2010; Veríssimo *et al.*, 2011; Broennimann *et al.*, 2012; Mourier *et al.*, 2012). The real challenge for any distributional study is determining which factors are responsible for the distribution of the study-species (Mourier *et al.*, 2012). This being said, competition has been shown to be a fundamental structuring force in communities, as it influences the realized niche of species which in turn facilitates community structure (Dulvy *et al.*, 2000; Krebs, 2001; Papastamatiou *et al.*, 2006; Fairclough *et al.*, 2008). Competition is, however, a negative-negative species interaction, by which members of the same species (intraspecific competition) or different species (interspecific competition) compete for the same limiting resources within their environment. Competition only takes place when resources are limiting, and there is a niche (spatial and/or dietary) overlap (Colwell and Futuyma, 1971; Sale, 1974; Pianka, 1981; Macleod, 2005; Papastamatiou *et al.*, 2006; Papastamatiou, 2008).

As competition reduces the fitness of all involved it is best "avoided" (Pianka, 1981; Ross, 1986), and co-existence through the differential use of resources is promoted (Sale,

1974; Pianka, 1981; Ross, 1986; Platell *et al.*, 1998; Macleod, 2005; Pikitch *et al.*, 2005; Knip *et al.*, 2012). Papastamatiou *et al.* (2006) studied the distribution of the four most common carcharhinid sharks in Hawaii using catch data, and noted that three of the species exhibited a degree of intraspecific variability in habitat use, as differences in their distribution by size, depth and sex became apparent. In addition to this, interspecific variability in habitat use was also evident. Although all four species of shark were recorded throughout the study area, each species' abundance peaked in a different region. These results led to the conclusion that these sharks may be spatially partitioning their physical habitat, both intraspecifically and interspecifically, thereby facilitating coexistence. As competition has been shown to play a strong role in habitat selection by sharks, habitat may always be a means by which shark species compete (Pianka, 1981; Methratta and Link, 2007; Knip *et al.*, 2012), and to facilitate co-existence and reduce the intensity of competition, sharks may divide their physical habitat on a spatial and temporal scale by location, depth and even habitat types (Simpfendorfer *et al.*, 2005; Flammang *et al.*, 2011). Thus, the manner in which shark species compete may result in the distributions and population structures currently observed (Connell, 1983; Fairclough *et al.*, 2008; Ruocco *et al.*, 2012).

As noted earlier, *Squalus acutipinnis* and *S. bassi* are common and co-occurring dogfish sharks, caught as bycatch off the West and South coasts of South Africa (Heemstra and Heemstra, 2004; Petersen *et al.*, 2008). They are both deep-sea demersal sharks that are similar in morphology (Heemstra and Heemstra, 2004; Ebert and Stehmann, 2013), are presumed to have similar resource requirements and may be potential competitors (Platell *et al.*, 1998; White *et al.*, 2004; Bethea *et al.*, 2007; Bizzarro *et al.*, 2007; Veríssimo *et al.*, 2011; Bornatowski *et al.*, 2014). In other parts of the world, the majority of dogfish are reported to have rather complex distribution structures (Braccini *et al.*, 2006a), with segregation in space and time being observed by maturity, size and sex (Hanchet, 1988; Compagno, 1990a; Graham, 2005; Braccini *et al.*, 2006a; Flammang *et al.*, 2011). This study, therefore, aims to determine the spatial distributions of *S. acutipinnis* and *S. bassi*, around the West and South coasts of South Africa, using the annual hake biomass trawl data collected by the national Department of Forestry, Fisheries and the Environment (DFFE). This study will furthermore, investigate the potential intraspecific variability in habitat use by each species, as well as assess the likely interspecific variability in habitat use between species, being presupposed that morphologically similar species that occur within the same environment are likely to be competitors (White *et al.*, 2004). It was hypothesized that:

23

Ho: There is no intraspecific difference in the size distribution of each species, with changing bathymetry on the West and South coast

Ho: There is no intraspecific difference in the sex distribution of each species with changing bathymetry on the West and South coast

Ho: There is no intraspecific difference in the size distribution, with sex, of each species on the West and South coast

Ho: There is no interspecific difference in the size distribution, with changing bathymetry on the West and South coast

Ho: There is no interspecific difference in the sex distribution with changing bathymetry on the West and South coast

Ho: There is no interspecific difference in the size distribution, with sex, on the West and South coast

Materials and methods

Study survey area and sampling design

All dog shark data samples were recorded/collected during routine demersal hake biomass surveys conducted by DFFE, on the West and South coasts of South Africa, between 1983 and 2015. Surveys extended from the Orange River (OR) to Port Elizabeth (PE, Fig. 1). Prior to 2010, the offshore limit of the survey area was based on the 500 m isobath, but the survey area was subsequently extended to the 1000 m isobath.

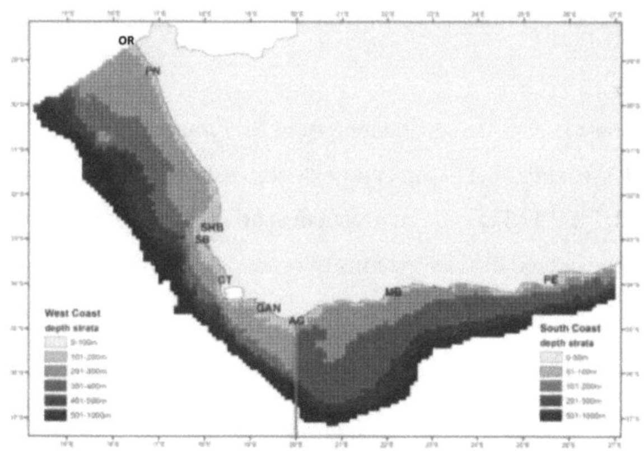

Figure 1: Survey grid used for demersal surveys conducted by DFFE. Grid cells are 5x5 minutes and are grouped into depth strata. The 20 °E meridian (Cape Agulhas) is the boundary between the West and South Coast surveys. PN=Port Nolloth, SHB=St. Helena Bay, SB=Saldana Bay, CT= Cape Town, GAN= Gansbaai, AG= Cape Agulhas, mB=Mossel Bay, PE=Port Elizabeth. Figure adapted from van der Heever 2017.

All trawls were conducted during daylight hours. Station selection followed a pseudorandom stratified sampling design, as explained by Payne *et al.* (1985). In summary: strata were defined by depth and latitude (West coast) or depth and longitude (South coast); the number of stations per stratum was directly proportional to the area of that stratum, and a proximity rule was applied to limit clumping of randomly selected positions. All trawls were intended to last up to a maximum of 30 min, but if the topography of the seabed was irregular and rough, trawls lasted for a shorter time. All West coast surveys were carried out annually, in summer (January and February), except during 1998 and 2012, when no surveys took place. All South coast surveys took place bi-annually, during autumn (March, April, May and June) and spring (September and October). No autumn surveys were conducted in 1998, 2002, 2012 and 2013, and no spring surveys were conducted from 1996 to 2002, 2005, and between 2009 and 2015.

Until 2003, the trawl gear consisted of a 2-panel 180-foot German trawl net with a rope-wrapped chain footrope and 50 m sweeps, 32 mm diameter trawl warp and 1.5 t WV otter boards (old gear). This was subsequently changed to a 4-panel 180-foot German trawl net with a modified rockhopper footrope and 9 m sweeps, 28 mm diameter trawl warps and 1.5 t Multipurpose Mogere otter boards (new gear). A higher net opening of this new gear allows for more pelagic catches and the shorter sweeps reduce the effect of herding (Leslie RW, *pers comm.*, 2019, DFFE; van der Heever, 2017). During 2006 and 2010, the old gear was used again in order to assist with the inter-calibration of the gears, as well as to provide overlap between the time series with the two gear types (Leslie *et al.*, 2013; van der Heever, 2017).

The West coast is considered to extend from the Orange River (29° 30' S) to Cape Agulhas (20° 00' E), whereas the South coast extends eastward from Cape Agulhas (20° E) to Port Elizabeth (25° 34' E; Fig. 1). Environments east and west of the 20° 00' E meridian at Cape Agulhas are considered distinct in terms of oceanography (Hutchings *et al.*, 2009; van der Lingen and Miller, 2014).

Data analyses

In this chapter, patterns of habitat use were inferred from depth utilization. Once the trawl net was brought aboard the vessel, all sharks were counted, sexed, measured (TL, cm) and weighed; total weight per species per catch was determined and data expressed as density (number of sharks per square nautical mile; as Eq 1).

$$Density = \frac{number\ caught}{area\ swept\ by\ the\ trawl\ net\ (nm^2)} \qquad\qquad Eq\ 1$$

In assessing patterns in the distribution of the two squaloid species, data were analysed by geographic region, depth, size and sex. Data were grouped into 100 m depth bins (0 m – 100 m; 101 m – 200 m; etc.), and depth bins are referenced by the midpoint of the depth bin (50 m, 150 m, etc.) in the text. Length data were grouped by size class, and considered the length at birth as well as the length at 50% maturity for *Squalus acutipinnis* (Lucifora *et al.*, 1999; Watson and Smale, 1999). Three size classes were established, namely small, medium and large. The "small" class, grouped sharks that measured $\leq$ 30 cm in total length (TL), the "medium" class, grouped sharks that measured between 31 cm and 39 cm in TL, while the "large" class grouped sharks that measured $\geq$ 40 cm in TL. Although *S. bassi* matures at a greater average size and attains a greater maximum size than *S. acutipinnis*, the same size categories were used for both species on the assumption that sharks of the same size would have similar resource requirements regardless of differences in maturity. Size classes do not relate to maturity levels.

In order to examine spatial patterns in the variously aggregated data, Quantum GIS was used. Depth frequency diagrams detailing the average number of sharks caught were plotted for each species, coast, and sex. This was calculated as the number of individuals (per species and size classs) per trawl, summed over all trawls (per coast and depth bin) and then divided by the number of trawls. Non-parametric multivariate Mann-Whitney U tests were then used to determine whether there were significant differences in the sizes (TL) of *Squalus acutipinnis* and *S. bassi* sharks sampled between the two coasts. Due to the fact that the TL data of both species were not normally distributed, non-parametric tests were done. Intraspecific and interspecific variability in size distribution patterns, by depth and along each coast were evaluated by means of Chi-square (χ^2) contingency tables (Zar, 1999, 2010). All data (male, female and unsexed) were collectively used in the analyses when overall results were provided, but data were also analysed separately by male and female.

The average density of males and females were calculated at each depth bin for each coast, and sex ratios were calculated. Kruskal Wallis tests were then used to evaluate intraspecific and interspecific differences in density for each coast based on shark size and depth of occurrence. Significance was tested at an alpha level of 0.05. Between 1983 and

2015, a total of 9 802 research survey trawls were conducted around South Africa by DFFE (Fig. 2), with 6 081 trawls being conducted on the West coast and 3 721 trawls on the South coast.

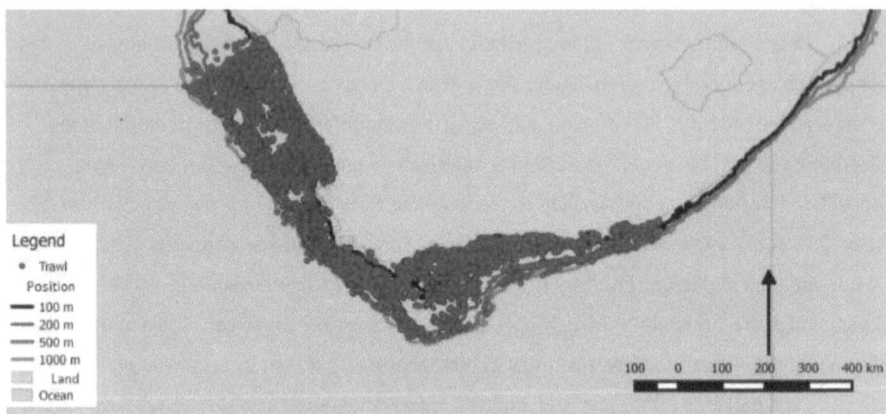

Figure 2: The positions of all demersal research trawls that have been conducted by the Department of Forestry, Fisheries and the Environment (DFFE) on their demersal hake biomass surveys from 1983 – 2015. The "gaps" in the distribution of the trawls are due to hard, rough ground that could not be sampled with the demersal trawl gear used during these surveys.

Results

***Intraspecific evaluation of habitat use by* Squalus acutipinnis**

Squalus acutipinnis were recorded in a total of 1 429 (14.6%) trawls; 259 (4.3%) on the West coast and 1 170 (31.4%) on the South coast (Fig. 3a). A total of 34 687 *S. acutipinnis* were caught, of which 15 539 were male, 17 215 were female and 1 933 could not be sexed. Sharks were caught between 24 m - 441 m on the South coast, and between 88 m - 407 m on the West coast. The results of the Mann-Whitney U test revealed significant differences in the sizes (TL) of *S. acutipinnis* between the coasts (U= 65.332; p <0.001). The majority of sharks inhabited the 150 m depth binbin on both coasts (Fig. 3a, 3b and 3c), and catches were dominated by medium-sized individuals (Fig. 3b and 3c). The results of the Chi-square contingency tests revealed significant differences in the size compositions with depth on both coasts (West; χ^2= 371.595; df= 8; p <0.05; South; χ^2= 601.359; df= 8; p <0.05). By comparison with larger sharks, small individuals were more frequently sampled at shallow depths, and as the depth increased, the average size of the sharks caught increased. This was noted for both coasts (Table 1; Fig. 3b and 3c).

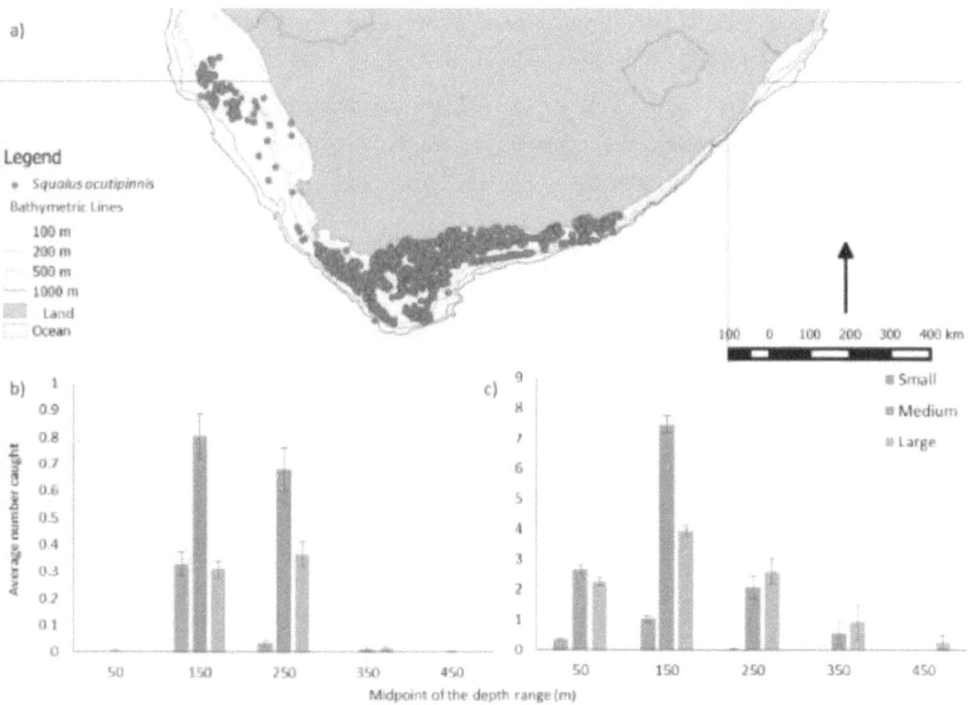

Figure 3: The latitudinal and bathymetric distribution (a) of the stations where *Squalus acutipinnis* were caught around the coast of South Africa during demersal research surveys conducted by the Department of Forestry, Fisheries and the Environment (DFFE) between 1983 and 2015. The average number of individuals caught (±SE) per size category and depth binbin per trawl is illustrated for the West (b) and South (c) coasts. Stations were placed into 100 m depth bins by the depth of capture. Depth bins are referenced by the midpoint (0 m – 100 m = 50 m; 101 m – 200 m = 150 m; etc.).

Table 1: The mean size (total length to the nearest cm below, ±SE) of *Squalus acutipinnis* caught per depth bin (m), along the West and South coasts of South Africa

	West coast	South coast
Depth bin (m)	cm	cm
0 - 100	42.5±0.01	46.6±0.52
101 - 200	39.8±0.73	44.3±0.36
201 - 300	44.2±0.87	48.8±1.20
301 - 400	48.3±1.36	53±3.45
401 - 500	42.5±0.01	66.7±0.87

Male *Squalus acutipinnis* were only sampled between 121 m – 407 m on the West coast, while males were recorded to occur between 24 m – 441 m on the South coast; and were most abundant in the 150 m depth binbins on both coasts (Fig. 4a, 4b and 4c). There were significant differences in the size composition of male *S. acutipinnis* caught at the different depths on both coasts (West: χ^2=136.737; *df* = 6; *p* <0.05; South: χ^2= 100.765; *df* = 6; *p* <0.05). Small males were more commonly recorded in shallow depths than other size classes and as the depth increased, the average size of the males caught increased (Table 2; Fig. 4b and 4c).

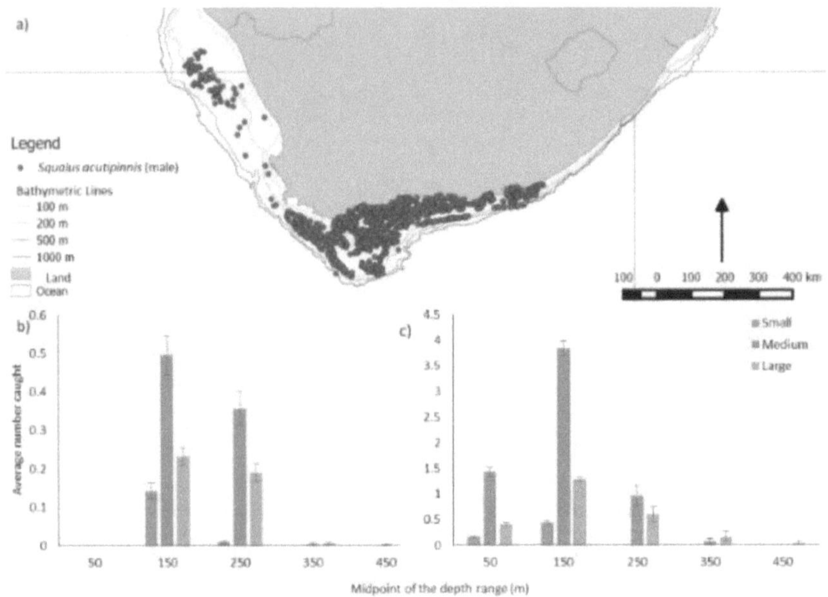

Figure 4: The latitudinal and bathymetric distribution (a) of the stations where male *Squalus acutipinnis* were caught around the coast of South Africa, during demersal research surveys conducted by the Department of Forestry, Fisheries and the Environment (DFFE) between 1983 and 2015. The average number of individuals caught (±SE) per size category and depth bin per trawl is illustrated for the West (b) and South (c) coasts. Stations were placed into 100 m depth bins by depth of capture. Depth bins are referenced by the midpoint (0 m - 100 m = 50 m; 101 m – 200 m = 150 m; etc.).

Female *Squalus acutipinnis* were caught between 88 m – 331 m on the West coast and between 24 m - 441 m on the South coast, and although most sharks were recorded in the 150 m depth binbin on the South coast, they peaked in abundance at both 150 m and 250 m on the West coast (Fig. 5a, 5b and 5c). As noted for the males, there were significant differences in the size composition of female *S. acutipinnis* caught at the different depths on both coasts (West: χ^2= 328.669; *df* = 6; *p* <0.05; South: χ^2= 564.955; *df* = 8; *p* <0.05). Small females were more commonly found in shallower depths than medium and large-sized females. As the depth increased, the average size of the females caught increased (Table 2; Fig. 5b and 5c).

Table 2: The mean size (to the nearest cm below) of male and female *Squalus acutipinnis* caught per depth bin (m), along the West and South coasts of South Africa

| | Male | | Female | |
| | West coast | South coast | West coast | South coast |
Depth bin (m)	cm	cm	cm	cm
0-100	-	40.5±0.29	42.5±0.01	47.6±0.40
101-200	39.6±0.40	40.7±0.19	36.1±0.54	44.4±0.26
201-300	43.8±0.54	44.6±0.95	44±0.63	48.7±0.87
301-400	51.1±0.35	48.1±0.96	51.1±0.01	48.6±0.94
401-500	42.5±0.01	72.5±0.01	-	65.8±0.87

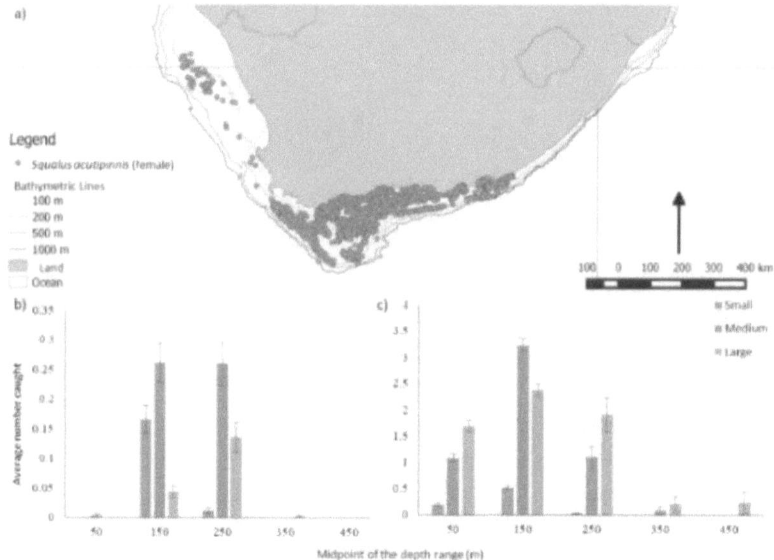

Figure 5: The latitudinal and bathymetric distribution (a) of the stations where female *Squalus acutipinnis* were caught around the coast of South Africa, during demersal research surveys conducted by the Department of Forestry, Fisheries and the Environment (DFFE) between 1983 and 2015. The average number of individuals caught (±SE) per size category and depth binbin per trawl is illustrated for the West (b) and South (c) coasts. Stations were placed into 100 m depth bins by depth of capture. Depth bins are referenced by the midpoint (0 m -100 m = 50 m; 101 m – 200 m = 150 m; etc.).

The distribution of males and females show overlap in geographic and bathymetric distributions (Fig. 6). There were significant differences in the size compositions of overlapping males and females at the 150 m and 250 m depth bins on the West coast and between 50 m, 150 m and 250 m depth bin on the South coast (Table 3). Males were better represented across size categories at these depths on the West coast while females were better represented on the South coast. Sex ratios indicated that males were more abundant than females on the West coast, while females were more abundant than males on the South coast (Table 4).

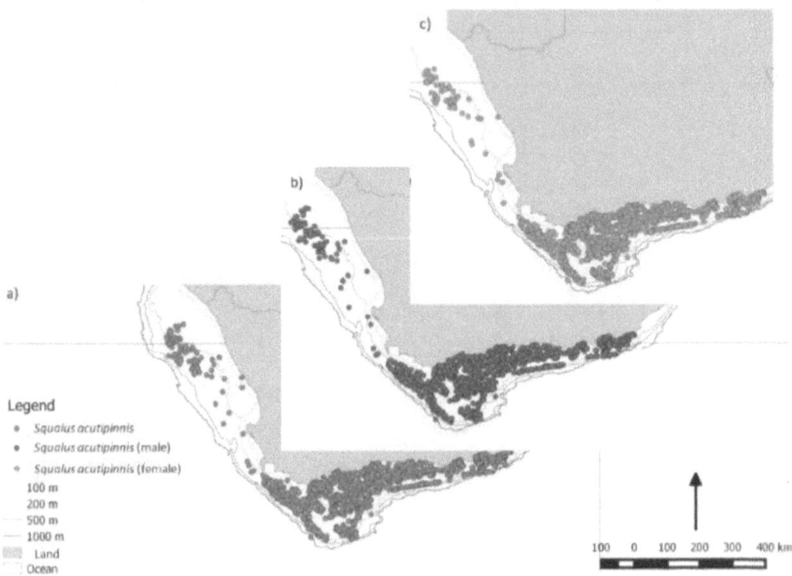

Figure 6: The (a) overlapping distribution of (b) male and (c) female *Squalus acutipinnis* around the West and South Coast of South Africa. Distribution is based on the catch data of male and female sharks which were provided by the Department of Forestry, Fisheries and the Environment (DFFE). Records were recorded on the hake biomass survey that was conducted by DFFE from 1983 – 2015.

Table 3: The chi-square (χ^2) values for the overall change in the size composition of overlapping male and female *Squalus acutipinnis*, with a change in bathymetric distribution, on the West and South coasts of South Africa. Significance is deemed at an alpha level of 0.05. Bold text indicates significant results and a dash indicates no result

Depth bin (m)	West coast			South coast		
	χ^2	*df*	*p*	χ^2	*df*	*p*
0-100	-	-	-	**687.06**	**2**	**<0.05**
101-200	**1425.69**	**2**	**<0.05**	**568.36**	**2**	**<0.05**
201-300	**768.13**	**2**	**<0.05**	**54.13**	**2**	**<0.05**
301-400	1.07	2	>0.05	0	2	>0.05
401-500	-	-	-	0	2	>0.05

Table 4: The ratio of the average density of male to that of female *Squalus acutipinnis* at various depth strata (m) along the West and South coasts of South Africa

Depth bin (m)	West coast Male:Female	South coast Male:Female
0-100	1.15:1	1:1.5
201-200	1.85:1	1:1.1
201-300	1.37:1	1:1.9
301-400	4.00:1	1:1.2
401-500	1.07:1	1:7.5

The results of the Kruskal Wallis (*H*) tests revealed significant differences in the overall density of *Squalus acutipinnis* across depth bin, on the West and South coasts (Table 5). Males displayed significant differences in the density distribution across depth bins, for each size category on both coasts, with the average number of small, medium and large males shown to peak at 150 m on both coasts (Table 6; Fig. 4b and 4c). Females, on the other hand, showed significant differences in density distribution across depth bins, for all size categories, except for medium-sized sharks on the West coast (Table 6). Moreover, the average number of small females peaked at 150 m, medium females at both 150 m and 250 m and large females at 250 m, on the West coast, while all three size categories peaked at 150 m on the South coast (Fig. 5b and 5c).

Table 5: The Kruskal Wallis (*H*) values for the density differences of small, medium and large *Squalus acutipinnis*, across depths, along the West and South coasts of South Africa. Significance is deemed at an alpha level of 0.05 and indicated in bold text

size class (cm)	West coast *H*	*df*	*p*	South Coast *H*	*df*	*P*
small	67.252	4	**<0.001**	55.593	4	**<0.001**
medium	26.425	4	**<0.001**	193.661	4	**<0.001**
large	17.258	4	**<0.001**	12.646	4	**<0.05**

Table 6: The Kruskal Wallis (*H*) values for the difference in density of small, medium and large male and females *Squalus acutipinnis*, across depths along the West and South coasts of South Africa. Significance is deemed at an alpha level of 0.05 and indicated in bold text

Size class (cm)		West coast small	medium	large	South coast small	medium	large
male	*H*	61.99	30.1	24.32	36.21	151.99	129.39
	df	3	3	3	4	4	4
	p	**<0.001**	**<0.001**	**<0.001**	**<0.001**	**<0.001**	**<0.001**
female	*H*	57.8	6.23	16.44	44.65	157.71	12.16
	df	3	3	3	4	4	4
	p	**<0.001**	>0.05	**<0.001**	**<0.001**	**<0.001**	**<0.05**

Intraspecific evaluation of habitat use by Squalus bassi

Squalus bassi were recorded in a total of 302 (3.1%) trawls; 188 (1.9%) on the West coast and 114 (1.2%) on the South coast (Fig. 7a). A total of 2 420 *Squalus bassi* were caught, of which 998 were male, 1 122 were female, and 300 were not sexed. Sharks were caught between 179 m – 722 m on the West coast, and between 103 m - 695 m on the South coast. The results of the Mann-Whitney U test revealed significant differences in the sizes (TL) of *S. bassi* between the coasts (U= 502.57; p <0.001). The majority of sharks inhabited the 250 m and 350 m depth bins on the West coast (Fig. 7b) and the 350 m, 450 m and 550 m depth bins on the South coast (Fig. 7c), and catches were dominated by large-sized individuals (Fig. 7b and 7c). The results of the chi-square contingency tests revealed significant differences in the size composition with depth on both coasts (West: χ^2= 177.801; df= 10; p <0.05; South: χ^2= 444.744; df= 10; p <0.05). Small individuals were more frequently sampled at shallow depths, and as the depths increased, the average size of sharks caught increased. This was noted for both coasts (Table 7; Fig. 7b and 7c).

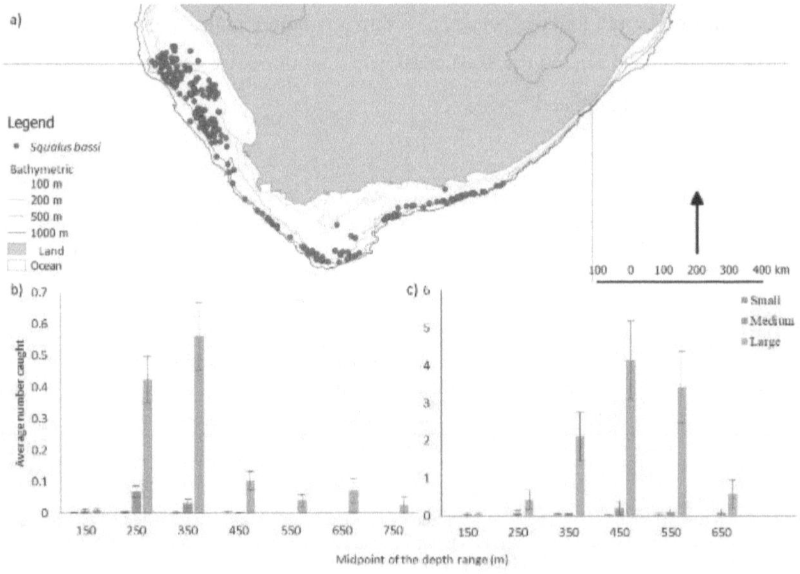

Figure 7: The latitudinal and bathymetric distribution (a) of the stations where *Squalus bassi* were caught around the coast of South Africa, during demersal research surveys conducted by the Department of Forestry, Fisheries and the Environment (DFFE) between 1983 and 2015. The average number of individuals caught (±SE) per size category and depth binbin per trawl is illustrated here for the West (b) and South coasts (c). Stations were placed into 100 m depth bins by depth of capture. Depth bins are referenced by the midpoint (0 m -100 m = 50 m; 101 m – 200 m = 150 m; etc.).

Table 7: The mean size (total length to the nearest cm below ±SE) of *Squalus bassi* caught per depth bin (m), along the West and South coasts of South Africa. A dash indicates no sharks were caught

	West coast	South coast
Depth bin (m)	cm	cm
0-100	-	-
101-200	58.6±1.05	59.8±0.39
201-300	59.0±0.96	65.4±0.45
301-400	63.9±0.96	70.2±1.88
401-500	70±0.71	72.1±1.73
501-600	76.9±0.40	74.3±1.14
601-700	82.1±0.36	74.7±0.30
701-800	97.5±0.01	-

Male *Squalus bassi* were only sampled between 189 m - 473 m on the West coast and 151 m – 695 m on the South coast. Males were most abundant in the 250 m and 350 m depth bins on the West coast (Fig. 8a and 8b) and in the 450 m depth bin on the South coast (Fig. 8a and 8c). There were significant differences in the size composition of male *S. bassi* caught at different depths on both coasts (West: $\chi^2 = 18.256$; $df = 6$; $p < 0.05$; South: $\chi^2 = 113.452$; $df = 10$; $p < 0.05$). Small males were more commonly recorded in shallow depths than other size classes and as the depth increases, the average size of the males caught increased (Table 8; Fig. 8b and 8c).

Table 8: The mean size (to the nearest cm below) of male and female *Squalus bassi* caught per depth bin (m), along the West and South Coasts of South Africa. A dash indicates no sharks were caught

	Male		Female	
	West coast	South coast	West coast	South coast
Depth bin (m)	cm	cm	cm	cm
0-100	-	-	-	-
101-200	62±0.56	48.5±1.91	36.1±0.27	58.4±1.44
201-300	59.3±0.58	61.9±1.68	54.7±0.92	73.6±1.88
301-400	61.3±0.56	70.8±0.75	64.8±0.84	69.4±1.81
401-500	68.8±0.30	67±0.99	79.1±0.18	72.4±1.57
501-600	-	71.2±0.49	79.5±0.38	74.8±1
601-700	-	74.2±0.24	82.1±0.36	75±0.22
701-800	-	-	67.5±0.01	-

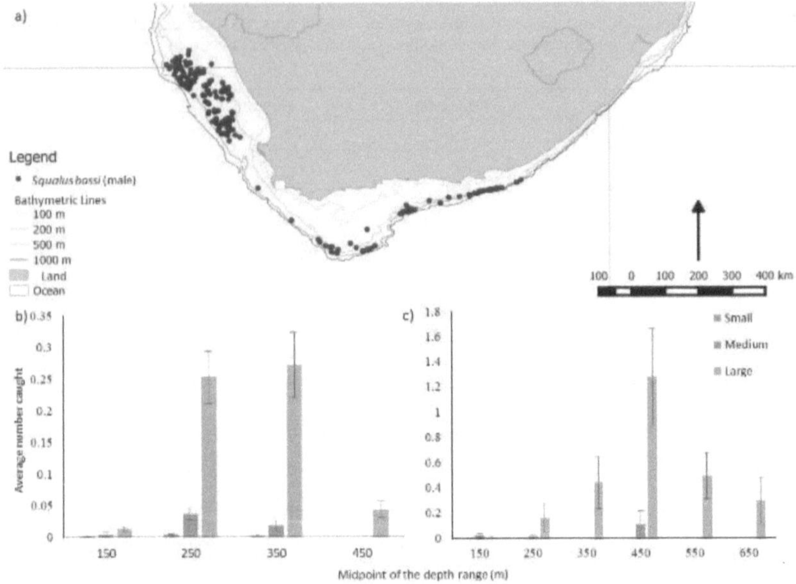

Figure 8: The latitudinal and bathymetric distribution (a) of the stations where male *Squalus bassi* were caught around the coast of South Africa, during demersal research surveys conducted by the Department of Forestry, Fisheries and the Environment (DFFE) between 1983 and 2015. The average number of individuals caught (±SE) per size category and depth binbin per trawl is illustrated here for the West (b) and South (c) coasts. Stations were placed into 100 m depth bins by depth of capture. Depth bins are referenced by the midpoint (0 m -100 m = 50 m; 101 m – 200 m = 150 m; etc.).

Female *Squalus bassi* were caught between 170 m - 722 m on the West coast and between 113 m – 695 m on the South coast. The abundance of females peaked at various depth binbins on both coasts, with abundances showing peaks in the 250 m and 350 m depth bins on the West coast (Fig. 9a and 9b), and in the 350 m, 450 m and 550 m depth binbins on the South coast (Fig. 9a and 9c). As noted for the males, there were significant differences in the size composition of female *S. bassi* caught at different depths on both coasts (West: χ^2=108.142; df= 12; p <0.05; South: χ^2=124.487; df= 10; p <0.05). Small females were more commonly found in shallower depths than medium and large-sized females. As the depth increased, the average size of the females caught increased (Table 8; Fig. 9b and 9c).

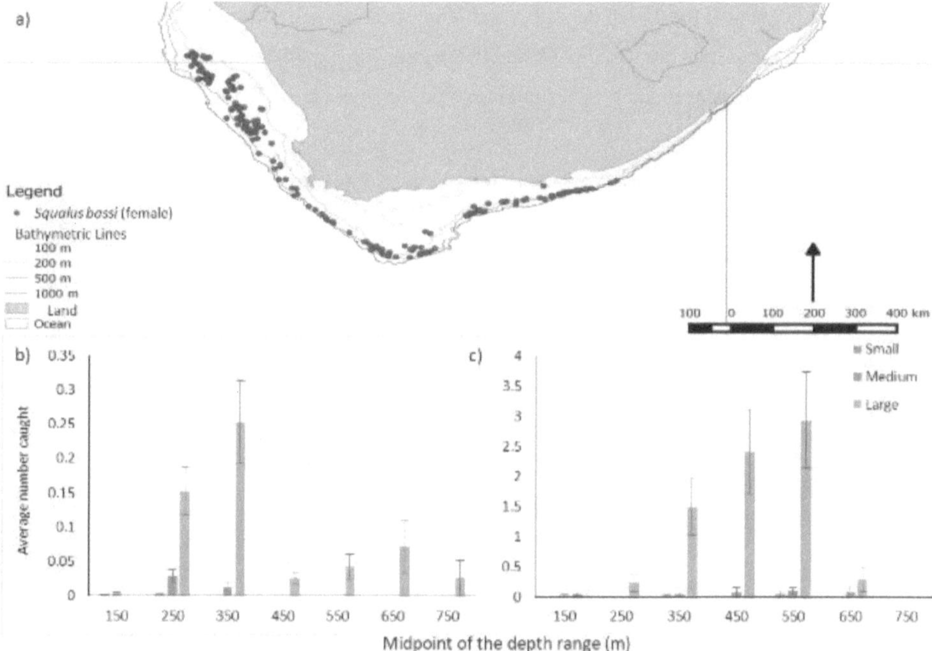

Figure 9: The latitudinal and bathymetric distribution (a) of the stations where female *Squalus bassi* were caught around the coast of South Africa, during demersal research surveys conducted by the Department of Forestry, Fisheries and the Environment (DFFE) between 1983 and 2015. The average number of individuals caught (±SE) per size category and depth binbin per trawl is illustrated here for the West (b) and South (c) coasts. Stations were placed into 100 m depth bins by depth of capture. Depth bins are referenced by the midpoint (0 m -100 m = 50 m; 101 m – 200 m = 150 m; etc.).

The distribution of males and females overlap in geographic and bathymetric distributions (Fig. 10). There were significant differences in the size compositions of overlapping males and females at the 150 m, 350 m and 450 m depth bins on the West coast; and only between individuals occurring in the 150 m depth bin on the South coast (Table 9). Males were better represented across size categories at these depth bins on the West coast while large females were better represented in the catches at 150 m on the South coast. Sex ratios indicated that males were more abundant than females on the West coast whereas females were more abundant than males on the South coast (Table 10).

The results of the Kruskal Wallis (*H*) test revealed significant differences in the overall density of *Squalus bassi*, across depth bins, on the West coast (Table 11). Males and females displayed significant differences in their density distribution across depths, for medium and large sharks on the West coast, with the average number of both medium males

36

and females peaking at the 250 m depth bin and large sharks at both 250 m and 350 m depth bins on the West coast (Table 12; Fig. 8b and 9b). Moreover, significant differences in the density distribution were found for medium males on the South coast (Table 12).

Table 10: The ratio of the average density of male to that of female *Squalus bassi* at various depth strata (m) along the West and South coasts of South Africa

	West coast	South coast
Depth bin (m)	Male:Female	
0-100	-	-
101-200	4.38:1	1:1.9
201-300	1.60:1	1:1.3
301-400	1.09:1	1:3.5
401-500	1.67:1	1:1.8
501-600	-	1:6.3
601-700	-	1:1.3
701-800	-	-

Table 11: The Kruskal Wallis (*H*) values for the density differences of small, medium and large *Squalus bassi*, across depth bins along the West and South coasts of South Africa. Significance is deemed at and an alpha level of 0.05 and indicated by bold text

	West coast			South coast		
Size class (cm)	*H*	*df*	*p*	*H*	*df*	*p*
small	14.36	6	**<0.05**	3.22	5	>0.0
medium	38.11	6	**<0.001**	10.45	5	>0.0
large	60.7	6	**<0.001**	10.06	5	>0.0

Table 12: The Kruskal Wallis (*H*) values for the difference in density of small, medium and large male and female *Squalus bassi*, across depth bins along the West and South coasts of South Africa. Significance is deemed at an alpha level of 0.05 and indicated by bold text. A dash indicates no result

		West coast			South coast		
		small	medium	large	small	medium	large
male	*H*	2.61	10.16	17.47	-	17.95	9.36
	df	3	3	3	-	6	6
	p	>0.05	**<0.05**	**<0.001**	-	**<0.05**	>0.05
female	*H*	-	23.09	14.13	4.56	5.42	4.13
	df	-	6	6	5	5	5
	p	-	**<0.001**	**<0.05**	>0.05	>0.05	>0.05

Interspecific distribution of Squalus acutipinnis *and* S. bassi

Even though *Squalus bassi* tends to occur offshore of *S. acutipinnis*, overlap zones in their distributions are evident at intermediate depths (Fig. 11). On the West coast, *S. acutipinnis* occurred offshore to 407 m while *S. bassi* occurred from 179 m – 722 m. On the South coast, *S. acutipinnis* occurred from inshore regions to 441 m, while *S. bassi* occurred from 103 m – 695 m. *Squalus acutipinnis* appears to have a relatively narrower bathymetric distribution, while *S. bassi* has a broad depth distribution. It should be remembered that *S. acutipinnis* were common in the trawls on the South coast while *S. bassi* were common in the trawls on the West coast.

Table 13: The chi-square (χ^2) values for the overall change in the size composition of overlapping *Squalus acutipinnis* and *S. bassi*, with a change in the bathymetric distribution along the West and South coasts of South Africa. Significance is deemed at an alpha level of 0.05 and indicated by bold text

Depth bin (m)	West coast			South coast		
	χ^2	*df*	*p*	χ^2	*df*	*p*
150	32.3	2	**<0.05**	32.83	2	**<0.05**
250	578.83	2	**<0.05**	32.29	2	**<0.05**
350	50.5	2	**<0.05**	61.79	2	**<0.05**
450	37.99	2	**<0.05**	0.88	2	>0.05

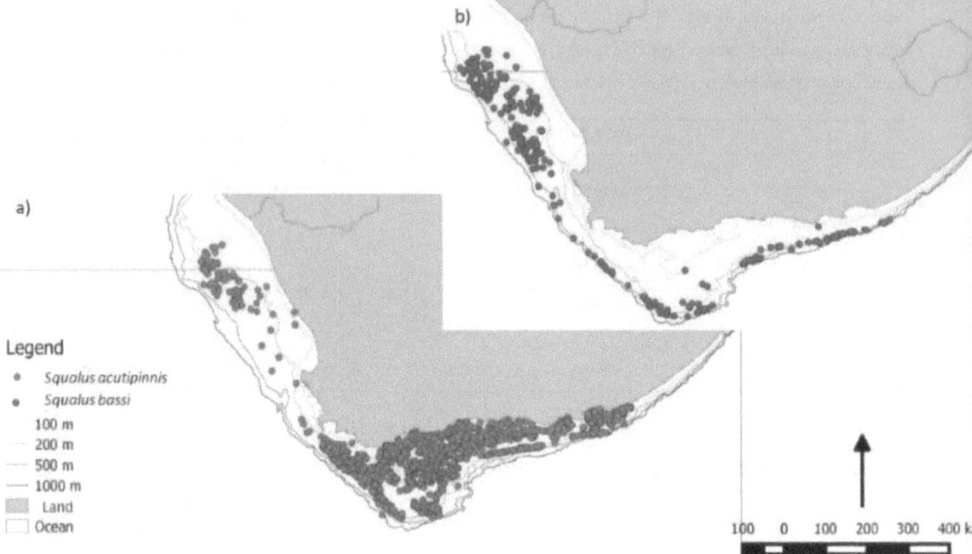

Figure 11: The overlapping distribution of a) *Squalus acutipinnis* and b) *S. bassi* around the West and South Coast of South Africa. Distribution is based on the catch data provided by the Department of Forestry, Fisheries and the Environment (DFFE). Catch data were recorded on the hake biomass surveys that were conducted by DFFE from 1983 – 2015.

here were significant differences in the size compositions of overlapping *Squalus acutipinnis* and *S. bassi* between 150 m – 450 m depth bins on the West coast, and between the 150 m – 350 m depth bins on the South coast (Table 13). Significant differences in the size compositions of overlapping males between 150 m – 450 m depth bins on the West coast and 250 m – 350 m depth bins on the South coast were detected using chi-square contingency tests (Table 14). Male *S. acutipinnis* dominated the catches (per average number individuals caught per trawl) across all three size categories, at 150 m and 250 m on the West coast and 250 m on the South coast. *Squalus bassi* males dominated catches at the 350 m and 450 m depth bins on the West coast and at the 350 m depth bin on the South coast. Moreover, females showed significant differences in size compositions at the 250 m depth bin on the West coast and between the 150 m – 350 m depth bins on the South coast (Table 14). Female *S. acutipinnis* dominated catches across all three size categories at 250 m on the West coast and 150 m and 250 m depth binson the South coast. Large female *S. bassi*, however, dominated the catches at 350 m on the South coast.

Table 14: The chi-square (χ^2) values for the overall change in the size composition of overlapping male and female *Squalus acutipinnis* and *S. bassi*, with a change in the bathymetric distribution along the West and South coasts of South Africa. Significance is deemed at an alpha level of 0.05 and indicated in bold text. A dash indicates no result

	Depth bin (m)	West coast			South coast		
		χ^2	*df*	*p*	χ^2	*df*	*p*
male	101-200	39.16	2	**<0.05**	3.92	2	>0.05
	201-300	342.61	2	**<0.05**	160.08	2	**<0.05**
	301-400	11.26	2	**<0.05**	10.48	2	**<0.05**
	401-500	31	2	**<0.05**	0.17	2	>0.05
female	101-200	1.5	2	>0.05	24.68	2	**<0.05**
	201-300	202.96	2	**<0.05**	25.56	2	**<0.05**
	301-400	0.1	2	>0.05	22.88	2	**<0.05**
	401-500	-	-	-	0.48	2	>0.05

Moreover, the density of each species caught differed at each depth (Fig. 12). Kruskal Wallis tests then revealed significant interspecific differences in the overall density of small ($H = 73.471$; $df = 1$; $p < 0.001$), medium ($H = 253.306$; $df = 1$; $p < 0.001$) and large ($H = 4.627$; $df = 1$; $p < 0.001$) sharks on the West coast. Small and medium *S. acutipinnis* had greater densities compared to *S. bassi* of similar sizes whereas the density of large *S. bassi* was greater than that of large *S. acutipinnis*. On the South coast, significant differences were only evident between small ($H = 53.318$; $df = 1$; $p < 0.001$) and medium ($H = 248.854$; $df = 1$; $p < 0.001$) sharks. Here, both small and medium *S. acutipinnis* displayed greater densities than *S. bassi*.

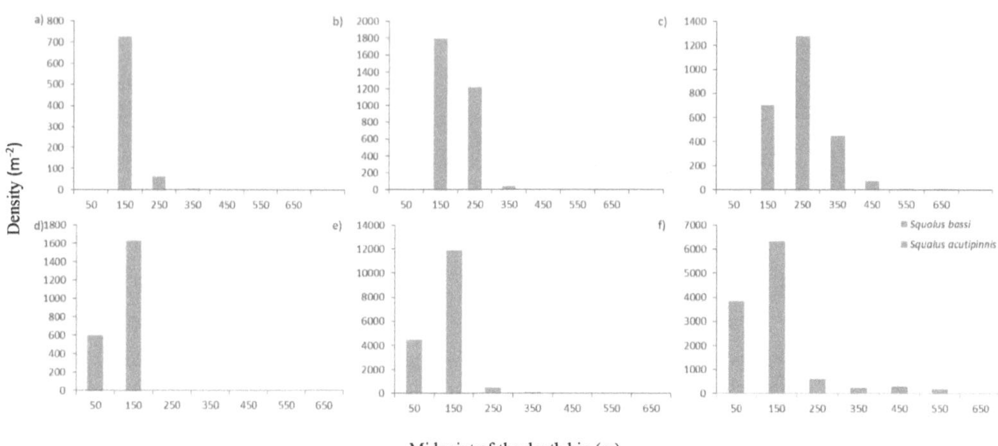

Figure 12: Average densities of small, medium and large *Squalus acutipinnis* and *S. bassi*, with depth (m), along the West (a, b, c) and South (d, e, f) coasts of South Africa. Stations were placed into 100 m depth bins by depth of capture. Depth bins are referenced by the midpoint (0 m -100 m = 50 m; 101 m – 200 m = 150 m; etc.).

Discussion

Squalus acutipinnis and *S. bassi* were recorded, variably, within their previously described distribution range (Heemstra and Heemstra, 2004). Both species were caught more frequently on the South coast than on the West coast, even though the majority of the trawling effort occurred on the West coast. It is difficult to unambiguously interpret this result because the two regions support quite different environments and habitat mixes (Lutjeharms, 2006; Hutchings *et al.*, 2009; van der Lingen and Miller, 2014). The South coast is characterised by a mix of warm Agulhas Current water and cooler upwelled water over the shelf; a pronounced seasonality in productivity (Lutjeharms, 2006; Hutchings *et al.*, 2009; van der Lingen and Miller, 2014) and a greater area of hard, untrawlable ground that may favour increases in squaloid abundance as noted too by Petersen *et al.* (2008).

Both species seem to be distributed more offshore on the West coast, while *Squalus acutipinnis* is shown to occur more inshore on the South coast. This could reflect the displacement of these species by other species with similar habits. Van der Heever (2017) examined the intraspecific and interspecific variability in the distribution of two catshark species, *Holohalaelurus regani* and *Scyliorhinus capensis,* around the coast of South Africa, finding an inverse relationship in their distributions: *H. regani* dominated catches on the West coast, while *S. capensis* dominated catches on the South coast. *Holohalaelurus regani* was also reported to occur at very high densities on the West coast (van der Heever, 2017) and, speculating, it is possible that this species displaces *S. acutipinnis* and *S. bassi* from the inshore region there. Carrassón *et al.* (1992) noted in the Catalan Sea that *Centroscymnus coelolepis* was strictly distributed between a depth bin of 1500 m – 2250 m, while in Japan it was confined to depths between 100 m – 1500 m. These authors concluded that *C. coelelepis* may have been displaced from its spatial niche to avoid competition with other species in Japanese waters. That said, these differences may also mirror changes in bottom substrate type, as habitats shallower than 200 m on the West coast mainly comprise sandy substrates (Leslie RW, *pers comm.*, 2019, DFFE), whereas a greater variety of bottom substrates are available on the South coast, which include numerous low and high profile reefs (Heyns *et al.*, 2016). *Squalus* species have previously been shown to occur in great abundances in untrawlable and mixed grounds (Petersen *et al.*, 2008; Worm *et al.*, 2013; Oliver *et al.*, 2015). The shallow waters off the northern parts of the West coast are also often oxygen-depleted and these sharks may simply be avoiding stressful low oxygen conditions (Butler and Taylor, 1975; Diaz and Rosenberg, 2008;

Speers-Roesch *et al.*, 2012; Zimmer and Wood, 2014).

Smaller *Squalus acutipinnis* and *S. bassi* both tended to inhabit shallower depth strata than their larger conspecifics and significant differences in the size composition of catches with depth were found. This indicates segregation by size and supports the "Bigger-Deeper" phenomenon displayed by many deep-sea sharks and teleost species, which is assumed to limit competition between different size and age classes (Kazunari and Tanaka, 1988; Macpherson and Duarte, 1991; Jakobsdóttir, 2001; Pikitch *et al.*, 2005; Oddone *et al.*, 2010). Moreover, the witnessed change in depth distribution may thus be attributed to the morphological changes that occur with size, with accompanying changes in diet requirements (Sims, 2003; Knip *et al.*, 2011; Espinoza *et al.*, 2012; van der Heever, 2017). Around the world, shallow waters serve as a nursery for younger sharks and are vital for the well-being of the shark populations (Sims *et al.*, 2001; Simpfendorfer *et al.*, 2005; Braccini *et al.*, 2006; Knip *et al.*, 2010). These "nursery" areas provide refuge to the young from predation (including by conspecifics) as well as providing small sharks with access to smaller prey that they are likely more capable of handling (Helfman, 1978, Sims *et al.*, 2001, Sims, 2003, Knip *et al.*, 2010) - shallow habitats and embayments being not only nursery habitats for small sharks, but many species of other fish too (Valesini *et al.*, 1997; White and Potter, 2004). Larger sharks, on the other hand, go deeper to increase their chances of obtaining larger prey that is likely more nutritious, and therefore, larger sharks are frequently caught in deeper waters and neonates in shallow waters (White and Potter, 2004; Simpfendorfer *et al.*, 2005; DeAngelis *et al.*, 2008).

Pikitch *et al.* (2005) examined the population structure and habitat use of elasmobranchs around a Caribbean atoll. Their results revealed size/age-specific partitioning of the habitat by *Carcharhinus perezi*, *Negaprion brevirostris* and *Ginglymostoma cirratum*. The differing patterns of distribution displayed by each species were concluded to be a reflection of the differential use habitats by the different shark sizes. Depth was the means by which individuals of different sizes partitioned their environment. The movement of larger sharks to the deeper waters may also be explained by their requirement for more nutritious meals (Methratta and Link, 2007). Habitat separation by depth segregation among sharks is therefore not an uncommon phenomenon as Papastamatiou *et al.* (2006) also described this for the degree of depth segregation displayed by immature and mature individuals of a variety of species around Hawaii.

Although it has previously been proposed that the Agulhas Bank (South coast) may be serving as a nursery habitat for *Squalus acutipinnis* (Watson and Smale, 1999), catches of small *Squalus acutipinnis* were still relatively scarce, with catches of small S. *bassi* even scarcer still. It is unlikely that some individuals escaped through the cod end of the trawl while others were caught (Graham, 2005; Graham and Daley, 2011). The rarity of small individuals in catches is not uncommon when trawling is the method of data collection (Braccini *et al.*, 2006a). This is because small (juvenile) dogfish have been said to be either mesopelagic or benthopelagic, thus spending most of their time feeding in the water column (Braccini *et al.*, 2006b; Graham and Daley, 2011). Indeed, Compagno *et al.* (1991) considered the then, small *S. megalops* (now *S. acutipinnis*) to be bathypelagic.

Caution should perhaps be exercised in interpreting the data for small *Squalus bassi* similarly, as their rarity may reflect a bias imposed by the sampling gear used (Wilson and Seki, 1994; Fischer *et al.*, 2006; Oddone *et al.*, 2010): skippers are reluctant to trawl on rough ground. For example, *Squalus mitsukurii* was the most abundant elasmobranch and consequently the most common bycatch species from 1985 – 1988 in the south-east Pacific Ocean, Hancock seamount (Wilson and Seki, 1994; Fischer *et al.*, 2006; Simpfendorfer and Kyne, 2009; Oddone *et al.*, 2010). During its "prime time," and before significant declines in the catch, there was a scarcity of small-sized *S. mitsukurii*, which these authors related to gear selectivity (Oddone *et al.*, 2010). The same could, therefore, apply to this study.

Segregation by sex is another common phenomenon amongst sharks, especially deep-sea species such as *Squalus* (Springer, 1967; Kazunari and Tanaka, 1983; Klimley, 1987; Sims *et al.*, 2001; Sims, 2003; Braccini *et al.*, 2006; Oddone *et al.*, 2010; Flammang *et al.*, 2011), with some species displaying sex-specific habitat requirements (Graham, 2005; Braccini *et al.*, 2006a; Pajuelo *et al.*, 2011; Jacoby *et al.*, 2012; Bangley and Rulifson, 2014). Within both *S. acutipinnis* and *S. bassi* species, differences in the sex composition of sharks at various depths were observed and it is not unlikely that sex is another means by which these species are segregating their environment. Reports of size and sexual segregation have been documented for *S. acutipinnis* in South Africa (Compagno, 1990a; Ebert and Stehman, 2013), and for *S. megalops* in New South Wales, where larger females segregate from juveniles and males (Graham, 2005). A study on *Squalus acanthias* off New Zealand (Ketchen, 1986), concluded that this species has an intricate population structure that is directly related to its reproductive cycles. Although ovulation, parturition, and mating occur in deep water, females spend their first year of pregnancy in shallower water (Ketchen, 1986; Hanchet, 1988; Braccini *et al.*, 2006a). During their second year of pregnancy, these females then move to deeper waters (Ketchen, 1986; Hanchet, 1988; Braccini *et al.*, 2006a). It was suggested that shallower waters are warmer, which helps speed up embryonic development (Ketchen, 1986; Hanchet, 1988; Braccini *et al.*, 2006a). Females are also have said to take refuge in shallow waters to avoid constant pursuit by males in the deeper water, as mating is exhausting and there is less competition in these waters (Ketchen, 1986; Hanchet, 1988; Braccini *et al.*, 2006a). Another example of such intricate behaviour is exhibited by *Squalus megalops* off southeastern Australia (Braccini *et al.*, 2006), where small male and female sharks segregate from the larger conspecific females, and females further segregated from one another depending on their term of gestation (Braccini *et al.*, 2006). Female *S. megalops* that are in their first year of pregnancy were found separated from those in their second year of pregnancy (Braccini *et al.*, 2006). This scenario may be what is occurring along the West and South coasts, as large, pregnant *S. acutipinnis* and *S. bassi* were found to occur in shallow and deep waters (Leslie RW, *pers comm.*, 2019, DFFE).

Intraspecific and interspecific habitat separation by *Squalus acutipinnis* and *S. bassi* was inferred from the combination of catch and density data collected along the West and South coasts. Habitat partitioning seems to have occurred by coast, depth of capture and size class. *Squalus bassi* was more common in the trawls on the West coast than *S. acutipinnis* and *visa-versa* for the South coast. As previously indicated, species that are similar in morphology and resource requirements may partition their environment, to reduce the effects of interspecific competition, but only if resources are limited (White *et al.*, 2004; Papastamatiou *et al.*, 2006). As a further example, resource separation among four species of Urolophidae was examined by Platell *et al.* (1998) in the coastal waters of Australia. Their results indicated that the abundance of each species was influenced by the abundance of the other (Platell *et al.*, 1998). Habitat separation was achieved by one species being more abundant in one habitat along the coast, but less abundant in the other regions where other batoids were found, thus reducing the potential of interspecific competition (Platell *et al.*, 1998). In a study of *Denia crepidalbus* and *D. calceus* from the Southeast Atlantic (Namibia), Kazunari, (1991) noted that where the two species co-occurred, they segregated vertically. By separating and segregating the habitat by depth, co-existence was achieved. When *S. acutipinnis* and *S. bassi* co-occur, habitat separation may be achieved by the *bro*ader geographical range but narrower depth bin of the former; and vice versa for *S. bassi.*

It should be remembered, however, that factors such as the physical and chemical parameters of the water column and physiological development may also contribute to the size-based population structure observed (Quigley, 1985; Butler and Taylor, 1975; Hopkins and Cech, 2003; Wiley and Simpfendorfer, 2007; Bizzarro *et al.*, 2014). After all, while significant differences in depth distributions by coast, size and sex were noted (both intraspecifically and interspecifically), there was still overlap. This overlap may have only been possible since resources were not limiting at that depth or because different sizes/sex fed on different foods available at that depth (Bethea *et al.*, 2004; Vaudo and Heithaus, 2011; van der Heever, 2017). Van der Heever, (2017) noted a similar pattern and also alluded to the above explanation.

Furthermore, two noticable features in the overall broad distribution patterns of the two species were the "breaks" in the distribution of both male and female *S. acutipinnis* on the West coast, as well the "break" in the distribution of male *S. bassi* on the West coast, which is not apparent in the distribution of the females. The break in male and female *S. acutipinnis* distributions may be as a result of species displacement by other shark species, whereas the break in the distribution of male *S. bassi* may be as a result of displacement by female *S. bassi* as well as other shark species. These reasons are however largely speculative and should definitely be noted as a topic for future research.

This study was the first step in assessing the distribution patterns and population structure of *Squalus acutipinnis* and *S. bassi*, based on their trawl footprint, around the coast of South Africa. The results highlighted the fact that *S. acutipinnis* and *S. bassi* are not randomly distributed but rather distributed in a structured way and may possibly be partitioning their physical habitat both intraspecifically and interspecifically, by coast, sex,depth and size, in order to facilitate co-existence. Distribution patterns were shown to be size- based and were inferred to be as a result of intraspecific and interspecific competition; ontogenetic changes with growth, habitat and dietary requirements as well as a result of reduced predation risk. Although this study has provided useful information on the distribution and demography of these species, and an understanding of these distribution patterns by sex, and size, coupling information on the diet will enhance these results as it will provide further insight into the possible effects that competition for food has on the observed distributions. Together, this may be important in developing changes to the fishing strategies that may mitigate the impact of trawling on the populations of these species.

Chapter Three

The dietary habits of two sympatric squaloid sharks, *Squalus acutipinnis* and *Squalus bassi*, off the South and West coasts of South Africa

Introduction

Knowledge on the feeding ecology of a species is important as it provides information on the role that a species plays within its environment (Wetherbee *et al.*, 1990; Vaudo and Heithaus, 2011; Espinoza *et al.*, 2012). The influence of habitat on the distribution of species has been shown to vary markedly, within and between species but it is the diet that is generally considered as the major reason for the observed distributions of most species (Ross, 1986; Badenhorst and Smale, 1991; Platell *et al.*, 1998; Hyndes *et al.*, 1999). The distribution of a predator, therefore, relates quite strongly to the distribution of its prey, and the local availability of prey within the preferred habitat of the predator will have an influence on the distribution of the predator within its preferred habitat (Heithaus, 2001).

As outlined previously, mesopredatory sharks serve as mid-trophic-level predators and play an important role in linking the lower and upper trophic levels and so play vital roles in ecosystem dynamics (Ritchie and Johnson, 2009; Tilley, 2011; Vaudo and Heithaus, 2011; Tilley *et al.*, 2013). The elucidation of trophic interactions of these sharks may help interpret changes in the environment with time, as well as the consequences thereof (Wetherbee *et al.*, 1990; Belleggia *et al.*, 2012). The latter is particularly important as we move towards an ecosystem approach to fisheries management (Cochrane *et al.*, 2004; Pikitch *et al.*, 2005; Mulas *et al.*, 2011; Valls *et al.*, 2011; López-García *et al.*, 2012; Wetherbee *et al.*, 2012; Bigman, 2013; Petersen *et al.*, 2015).

Unfortunately, due to increased fishing pressure and declining catches in shallow waters, fishing fleets have started fishing further offshore and in the deeper waters where these mesopredators are found in abundance (Ferretti *et al.*, 2008, 2010; Simpfendorfer and Kyne, 2009; Norse *et al.*, 2012; Schoener *et al.*, 2012). Around the world, declines in the top predators as a result of unsustainable fishing practices have been reported to have cascading effects on ecosystems, but yet there is little literature available that focuses on the impacts of fishing on the mesopredator sharks of the deep (Stevens *et al.*, 2000; Ferretti *et al.*, 2008,

2013; Heithaus *et al.*, 2008; Dulvy *et al.*, 2014; Worm *et al.*, 2013). Despite their importance in ecosystem stability, studies on the feeding ecology of these mesopredators are relatively scarce. This contrasts with the situation for larger and more charismatic species (Wetherbee *et al.*, 1990, 2012; Bush, 2003; Hoffmayer and Parsons, 2003; Bethea *et al.*, 2004; Papastamatiou *et al.*, 2006), which limits our understanding of their ecological role and impact on marine ecosystem (Wetherbee *et al.*, 1990, 2012; Hoffmayer and Parsons, 2003).

Although *Squalus acutipinnis* and *S. bassi* are common around South Africa, information on the diet of these species is limited to a single study conducted 28 years ago, the results of which implied differential resource-use (Ebert *et al.*, 1992). Both species fed on the same prey groups (teleosts, cephalopods and to a lesser extent, polychaetes and crustaceans) but there were differences in prey species (Ebert *et al.*, 1992). *Squalus acutipinnis* fed mainly on myctophids and *Octopus vulgaris*, whereas *S. bassi* fed predominantly on hake and *Todarodes angolensis* (Ebert *et al.*, 1992).

When closely related species occur in sympatry, differences in their diet may become apparent (Schoener, 1974; Platell *et al.*, 1998; Bethea *et al.*, 2004, 2007; Papastamatiou *et al.*, 2006; Papastamatiou, 2008; Yick *et al.*, 2011; Dunn *et al.*, 2013; de Sousa Rangel *et al.*, 2019). It has been suggested that these differences have evolved to allow the species to separate or rather divide their physical environment and/or resources, thereby allowing co-existence (Schoener, 1974). *Squalus acutipinnis* and *S. bassi* are morphologically and biologically similar, closely related species which have been shown to overlap in latitudinal and bathymetric distributions (see Chapter Two). Their co-existence may be possible due to their differential use of resources within their environment. This chapter aims to describe the overall diet of both species and evaluate resource-use, by coast, sex, depth and size class. Resource separation by these potentially competing species will be inferred based on indirect evidence, viz. their diets as estimated by stomach contents. It was hypothesized that:

H_0: There is no difference in the intraspecific resource-use, of each species, by coast, sex, size and depth

H_0: There is no interspecific difference in resource-use, between the species, by coast, sex, size and depth

Materials and methods

Study area and sampling design

All stomach samples were collected during the 1984 – 1994 and 2014 – 2016 routine demersal hake biomass surveys conducted by DFFE on the West and South coasts of South Africa (Fig. 1). All stomach samples collected during 1984 – 1994, were collected on the South coast whereas stomach samples collected during the 2014 - 2016 surveys constituted samples that were collected on both the West and South coasts. An explanation of the study area as well as survey design follows Payne *et al.* (1985) and has been described in Chapter Two.

Sample collection and laboratory analysis

As the trawl net was brought aboard the vessel, all sharks were sexed, counted, measured (TL, cm) and weighed as described in Chapter Two. If fish were collected alive, they were then immediately returned to the sea, but if dead then between 1984 and 1994, the diet of all sharks was immediately assessed by stomach content analysis. Between 2014 and 2016, if catches were small then all animals were dissected and their stomach contents analysed, but if >25 specimens of the same species (approximate size and sex) were caught at the same station, the catch was sub-sampled and 10 sharks were chosen, at random.

A ventral incision was made through the abdominal wall from the head to the cloaca: the stomach was removed and separated from the duodenum and oesophagus, blotted dry and weighed (g). When possible, stomach contents were immediately analysed in the ships laboratory and prey items were identified to the lowest taxonomic level, counted and weighed. When it was not possible to process the fish at sea, whole specimens were frozen at -20°C and then analysed back at the laboratory, where they were thawed and the stomach contents fixed and preserved in 10% formalin for 48 hours. Thereafter, prey items were rinsed in water and preserved in 70% ethanol until examination. Identifiable cephalopod beaks and otoliths were used to identify species and determine size at ingestion when the whole prey was well-digested and unidentifiable.

Identification of prey

Identification of the different prey items found in the stomachs was undertaken using field guides, taxonomic keys and in some instances, expert advice, as prey items were at various stages of digestion. Six major prey categories were considered when identifying prey, namely polychaetes, crustaceans, cephalopods, teleosts, other and unidentified. The category, "other" contained all prey items that were rare in the combined diet of both species while the category "unidentified" contained all prey items that were in a digestion stage that made it impossible to identify. The diets of these sharks are all listed in appendices and grouped by phylum, except for unidentified taxa. The category 'Other" is made up of all other phyla not explained above while the unidentified semi-digested material category in the appendices comprise all unidentified taxa.

In order to examine relationships between squaloid length and prey size, Spearman Rank correlations were computed between TL and cephalopod beak length. Cephalopod beaks are hard and relatively indigestible, and they were viewed under a dissecting microscope and measured using a graticule at 16 x magnification, from which the total length of the mantle of the cephalopods could be determined using the linear and allometric relationships provided by Clarke, (1986). Cephalopod beaks can remain in the stomachs of predators for a relatively long period of time. Unfortunately, neither otoliths nor crustacean carapace fragments could be reliably used to infer original prey size, because the majority of otoliths were eroded and relationships to infer prey size from carapace fragments were not possible as fragments were heavily digested.

Data analyses

In order to accurately describe the overall diet of these sharks, the minimum number of stomachs required to do so was determined using cumulative prey curves, following Ferry and Calliet (1996). The cumulative prey curves depicted the cumulative number of stomach sampled, against the cumulative diversity of identified prey, for each species of each size class, on each coast. Cumulative prey curves are founded on the fact that as the number of stomach sampled increases, variation in prey species diversity decreases, causing the curve to reach an asymptote, as new prey types are being introduced very rarely into the diet (Ferry and Calliet, 2021). As it is unreliable to visually inspect prey curves for asymptotes, the slope of the linear regression (b), through the last five samples was used as an unbiased criteria, where b≤0.05 signified an acceptable leveling-off of the prey curve for the characterization of diet (Bizzarro *et al.*, 2009; Brown *et al.*, 2012). Stomach orders were randomized 1000 times before curves were plotted to avoid bias. The 1000 randomisations were done in PRIMER 6 and all regression graphs were computed in MSExcel. All stomachs that were empty upon dissection were excluded from these analyses. Data were grouped by coast, size and depth classes following Chapter Two.

The diet of each species, sex, on each coast, at each depth and size class, was quantified by three relative indices of prey importance namely, number (N), weight (W) and frequency of occurrence (FO), as described by Hyslop, (1980). Each index was reported by percentage i.e, *%N* (Eq 2), *%W* (Eq 3) and *%FO* (Eq 4; (Hyslop, 1980; Tirasin and Jørgensen, 1999; Bizzarro *et al.*, 2007; Braccini, 2008; Baker *et al.*, 2014)), and associated advantages and disadvantages are discussed below. These indices provide different insights into the feeding habit of these shark species.

$$\%N = 100 \times \frac{total\ number\ of\ individuals\ of\ the\ ith\ prey\ category}{total\ number\ of\ all\ prey\ across\ all\ prey\ categories} \qquad \text{Eq 2}$$

The percentage number (*%N*) of each prey item was calculated as shown above. This method has the advantage of being easy to calculate and provides information on the feeding behaviour of predators as it reflects the density dependant acquisition of prey by predators (Hyslop, 1980; Macdonald and Green, 1983; Tirasin and Jørgensen, 1999). On the other hand, it only "works" if all prey items are identifiable, which is not always the case. In addition to this, if small prey items are to occur in great numbers in the diet of predators, its importance may be over-emphasized (Hyslop, 1980).

$$\%W = 100 \times \frac{total\ weight\ of\ individuals\ of\ the\ ith\ prey\ category}{total\ weight\ of\ all\ prey\ across\ all\ prey\ categories} \qquad \text{Eq 3}$$

The percentage weight (*%W*) of each prey item was calculated as described above. The advantage of this index is that it is easy to calculate and provides a sense of the nutritional value of prey consumed (Hyslop, 1980; Macdonald and Green, 1983; Tirasin and Jørgensen, 1999). However, it has many disadvantages. Firstly, it will only work if all prey items are identifiable and weights are known, which is seldomly the case. It may also overplay the importance of slowly digested, rare and/or heavy prey items, which neglects the importance of the easily digestible, small and abundant prey. Furthermore, trapped water/preservation chemicals inside prey may also result in skewed prey weights (Hyslop, 1980; Braccini, 2008).

$$\%FO = 100 \times \frac{total\ number\ of\ stomachs\ containing\ the\ ith\ prey\ category}{total\ number\ of\ stomachs\ analysed} \qquad \text{Eq 4}$$

The percentage frequency of occurrence of prey items was calculated as shown above. This method differs from the previous two indices as it does not describe the diet of the predators but rather indicates the variability of prey species in the diet of the predator (Hyslop, 1980; Macdonald and Green, 1983; Tirasin and Jørgensen, 1999). It is easy to calculate and self-explanatory. All prey items or their remnants are counted and the level of digestion doesn't matter, as long as prey are identifiable. A disadvantage of this method is that it too favours slowly digested prey. In comparison with $\%N$ and $\%W$ however, this index provides the most robust index of dietary composition (Hyslop, 1980; Bizzarro *et al.*, 2007; Baker *et al.*, 2014).

As each index has its own bias, a compound index, the index of relative importance (*IRI*; Eq 5) was further calculated (Hyslop, 1980; Pianka, 1981; Cortés, 1997; Tirasin and Jørgensen, 1999; Bizzarro *et al.*, 2007).

$$IRI = \%FO \times (\%N + \%W) \qquad \text{Eq 5}$$

The *%IRI* (Eq 6) was then calculated as it allows for easy comparison with results from other studies (Cortés, 1997, 1998, 1999; Bizzarro *et al.*, 2007; Baker *et al.*, 2014).

$$\%IRI = 100 \times \frac{IRI\ of\ the\ ith\ category}{total\ IRI\ across\ all\ prey\ categories} \qquad \text{Eq 6}$$

The main disadvantage associated with this method relates to possible taxonomic discrepancies and differing resolutions of prey identification between studies (Bizzarro *et al.*, 2007; Fanelli *et al.*, 2009; Brown *et al.*, 2012; Baker *et al.*, 2014; Bonnici *et al.*, 2018).

Intraspecific and interspecific variability in the diet of both species were evaluated by sex, coast, size class and depth class. As most stomach samples contained only one prey item, the stomach samples were pooled by species, coast, sex, depth and size to overcome the problems associated with analysing stomachs containing just one prey species (Linke *et al.*, 2001; Platell and Potter, 2001).

Where possible, intraspecific and interspecific differences in the frequency of occurrence of prey items among the major prey categories were tested using Chi-square contingency tables (Zar, 2010): pools with less than five specimens were excluded. Tests compared stomach content data: 1) between coasts for sharks of the same size and depth of capture; 2) for each size class, on each coast, with changing depth; 3) and between the different size classes, caught at the same depth and on the same coast. Owing to the fact that the frequency of occurrence data was deemed the most robust (Baker *et al.*, 2014), this index was used for all analyses, whereas *%IRI* was used to interpret some results, (following Baker *et al.*, 2014). In any analysis where the sample sizes differed between the groups being compared, the diet data of the group with the greater number of stomach samples was randomised 99 times without replacement and subsampled, such that groups being compared had the same sample size. Significance was tested at an alpha level of 0.05, after adjusting for multiple testing using the Bonferroni correction (Cooper, 1968).

Results

The total number of stomachs collected and analysed for each species varied between coasts. Of a total of 159 stomachs of *Squalus acutipinnis* that were collected on the West coast, 34 (21.4%) were empty; whereas only seven (7.5%) of the 93 stomachs of *S. bassi* were empty (Table 15). On the other hand, of the 868 stomachs of *S. acutipinnis* that were collected on the South coast, 108 (12.4%) were empty, whereas 41 (13.1%) of the 313 stomachs of *S. bassi* collected were empty (Table 15). Overall, across the six different prey categories, 118 different prey items were observed, with teleosts, crustaceans and cephalopods being most common (Table 16). It is important to note that most of the samples collected between 1994 and 1998 were of *S. acutipinnis,* from the South coast.

Table 15: The number of *Squalus acutipinnis* and *S. bassi* stomachs analysed for prey items from the West and South coasts. Samples were collected during the 1984 – 1992 and 2014 – 2016 annual hake biomass surveys conducted by the Department of Forestry, Fisheries and the Environment (DFFE)

	Squalus acutipinnis			*Squalus bassi*		
Coast	Non-empty	Empty	Total	Non-empty	Empty	Total
West	125	34	159	86	7	93
South	760	108	868	272	41	313
Total	885	142	1027	358	48	406

Table 16: The number of different prey by prey category found in the overall diet of *Squalus acutipinnis* and *S. bassi* on the West and South coast of South Africa

Prey Category	Taxa
Polychaetes	4
Crustaceans	25
Cephalopods	25
Teleosts	46
Other	16
Unidentified semi-digested material	2

In general, the smallest and largest sharks of both species were caught on the South coast (Table 17). The mean TL (cm) for *Squalus bassi* was larger than that of *S. acutipinnis* for the size classes, medium and large on the West coast and only large on the South coast (Table 17). Small *S. bassi* were not sampled on either coast (Table 17). The majority of *S. acutipinnis* were sampled in the 150 m depth bin on both coasts, whereas most individuals collected of *S. bassi* were sampled in 250 m and 450 m depth bins on the West and South coast, respectively (Table 17). Statistical analyses were often difficult because the number of stomachs available for analysis (by size and depth class) varied (Table 17). Whilst all data pools with ≥ 5 stomach samples were analysed, special focus has been given to those samples from medium and large-sized individuals occurring in the 150 and 250 m depth bins, because they were most numerous (Table 17), allowing for inter-and intra-species comparisons. As noted in Chapter Two, the size of the sharks caught increased with increasing depth (Table 17).

Table 17: The total number of analysed *Squalus acutipinnis* and *S. bassi* stomachs, per size class and the midpoint of the depth bin (m) sampled, off the West and South coasts of South Africa. The mean total length (TL, cm; ±SE) and size range, for each species, across size classes are also presented. Samples were collected during the 1984 – 1994 and 2014 – 2016 annual hake biomass surveys conducted by the Department of Forestry, Fisheries and the Environment (DFFE)

| | West coast | | | | | | South coast | | | | | |
| | *Squalus acutipinnis* | | | *Squalus bassi* | | | *Squalus acutipinnis* | | | *Squalus bassi* | | |
m	S	M	L	S	M	L	S	M	L	S	M	L
50	0	10	12	0	0	0	26	84	214	0	0	0
150	2	21	35	0	3	14	35	159	228	0	24	57
250	7	12	20	0	6	34	4	0	7	0	6	31
350	0	0	6	0	0	13	0	0	0	0	0	0
450	0	0	0	0	1	1	0	0	0	0	6	87
550	0	0	0	0	0	12	0	0	3	0	0	55
650	0	0	0	0	0	2	0	0	0	0	0	6
Total	9	43	73	NA	10	76	65	243	452	NA	36	236
Mean TL	27.6	41	48.6	NA	42.6	68.2	26.8	40.3	56	NA	35	73.5
±SE	0.71	0.69	1.11	NA	0.62	1.62	0.26	0.26	0.32	NA	0.95	1.03
Size Range	24 – 76			39 – 96			21 – 88			30- 104		

Size Classes: small (S); medium (M); large (L)

Cumulative prey curves show that in the case of *Squalus acutipinnis* (Fig. 13), it was not possible to make robust descriptions of the diet for any size class on the West coast (Fig. 13a, 13c and 13e) or of small animals on the South coast (Fig.13b), because the cumulative prey curves were not asymptotic, indicating that insufficient numbers of samples had been collected. For *S. bassi*, not enough stomachs were collected to provide comprehensive information about the diet of medium and large sharks on the West coast, or of medium and large sharks on the South coast (Fig. 14a – d). No correlation was evident between the size of cephalopod prey and the size of *S. acutipinnis* or *S. bassi* (Fig. 15).

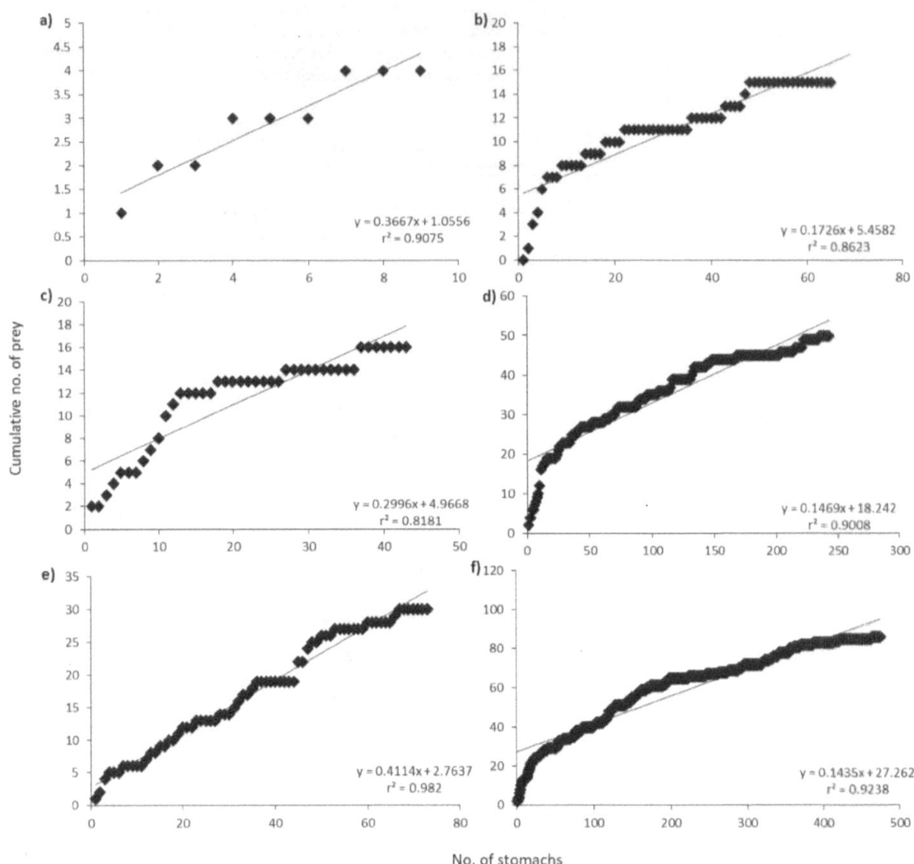

Figure 13: A randomised cumulative prey curve of the number of new prey taxa observed against the cumulative number of stomachs analysed by stomach content analyses of small (a; b), medium (c; d) and large (e; f) sized *Squalus acutipinnis* caught as bycatch off the West (a; c; e) and South (b; d; f) coasts of South Africa during 1984 – 1994 and 2014 – 2016 annual hake biomass surveys conducted by DFFE.

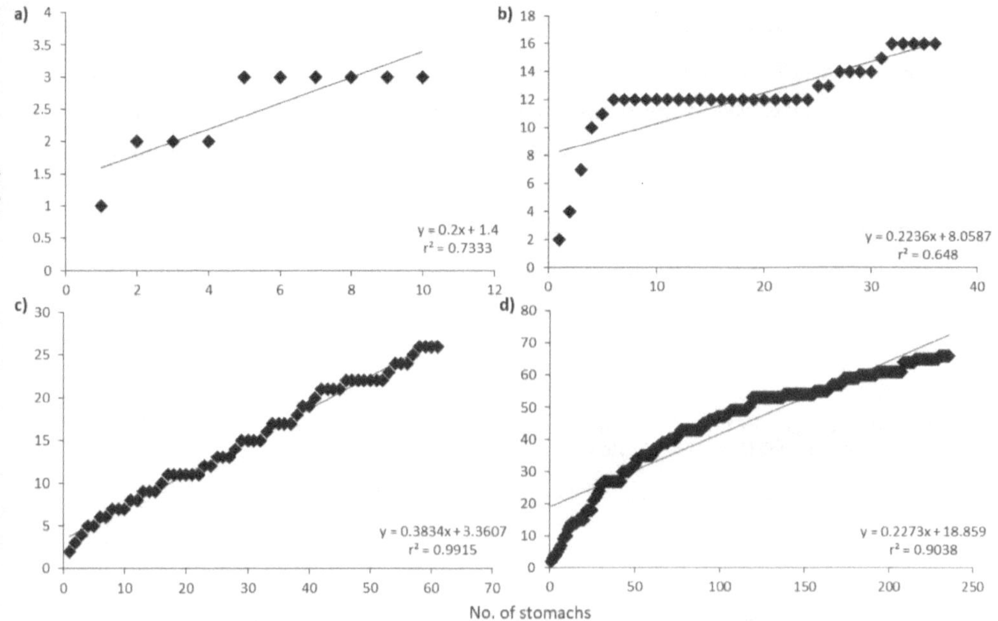

Figure 14: A randomised cumulative prey curve of the number of new prey items described against the cumulative number of stomachs analysed by stomach content analyses of medium (a; b) and large (c; d) sized *Squalus bassi* caught as bycatch off the West (a; c) and South (b; d) coast of South Africa during 1984 – 1994 and 2014 – 2014 annual hake biomass surveys conducted by DFFE.

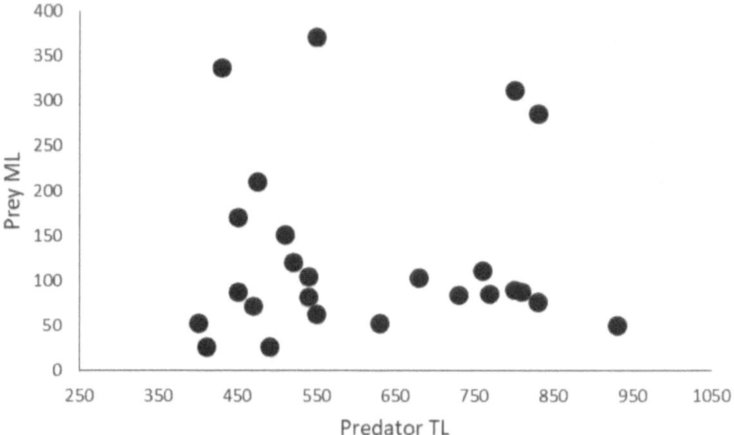

Figure 15: Change in cephalopod prey mantle length (ML; mm) with predator size (mm); r^2 = 0.114

Intraspecific evaluation of resource-use by **Squalus acutipinnis**

A total of 88 different prey taxa were found in the stomachs of *Squalus acutipinnis*: 32 on the West coast and 83 on the South coast. Teleosts were the most important prey category in the overall diet of these sharks, with crustaceans and cephalopods of lesser importance (Table 18). Dietary comparisons by sex could only be made between medium and large individuals captured in the 50 m and 150 m depth bins, owing to the paucity of data at other depths, and there was no apparent difference in the frequency of occurrence of prey categories in the diets of sharks by sex at either 50 m ($\chi^2 = 0.842$, $df = 3$; $p > 0.05$; N = 214) or 150 m ($\chi^2 = 7.442$, $df = 3$; $p > 0.05$; N = 228) depth bins.

Table 18: The number of different prey items by prey category found in the overall diet of *Squalus acutipinnis* on the West and South Coast of South Africa

Prey Categories	West coast	South coast
Polychaetes	1	2
Crustaceans	10	21
Cephalopods	4	18
Teleosts	15	28
Other	1	13
Unidentified semi-digested material	1	1

Small *Squalus acutipinnis* were caught in the 250 m (N= 7; Appendix 1) depth bin on the West coast and in the 50 m (N= 26; Appendix 2) and 150 m (N= 35; Appendix 3) depth bins on the South coast. Based on *%IRI*, teleosts (Appendix 1), polychaetes (Appendix 2) and crustaceans (Appendix 3) were the most important prey categories for sharks caught at each depth, respectively.

Medium *Squalus acutipinnis* were caught in the 50 m (N= 10; Appendix 4), 150 m (N= 21; Appendix 5) and 250 m (N= 12; Appendix 6) depth bins on the West coast, and in the 50 m (N= 84; Appendix 7) and 150 m (N= 159; Appendix 8) depth bins on the South coast. Polychaetes (Appendix 4) and teleosts (Appendix 5 and 6) were the most important prey categories in the diet of medium sharks caught on the West coast, whereas teleosts (Appendix 7 and 8) were most important on the South coast.

Large *Squalus acutipinnis* were caught at the 50 m (N= 12; Appendix 9), 150 m (N= 35; Appendix 10), 250 m (N= 20; Appendix 11) and 350 m (N= 6; Appendix 12) depth bins on the West coast and at the 50 m (N= 214; Appendix 13), 150 m (N= 228; Appendix 14) and

250 m (N= 7; Appendix 15) depth bins on the South coast. Based on the *%IRI*, the diet on the West coast was dominated by polychaetes in the 50 m depth bin (Appendix 9) and teleosts at greater depths (Appendix 10 -12), whereas the diet on the South coast was dominated by teleosts at all depths (Appendix 13 - 15). Appendices show the complete diet data from which subsampling may have or may not have taken place.

There were significant differences in the *%FO* of the different prey categories by coast for medium and large *Squalus acutipinnis* occurring at the 50 m (medium: $\chi^2 = 9.824$, $df = 3$, $p <0.05$; large: $\chi^2 = 34.762$ $df = 5$, $p <0.05$) and 150 m (medium: $\chi^2 = 17.713$, $df = 5$, $p <0.05$; large: $\chi^2 = 29.829$, $df = 5$, $p <0.05$) depth bins. Polychaetes were important in the diet of medium sharks at the 50 m depth bin on the West coast while teleosts were common in sharks at the same depth on the South coast (Fig. 16).

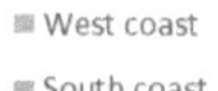
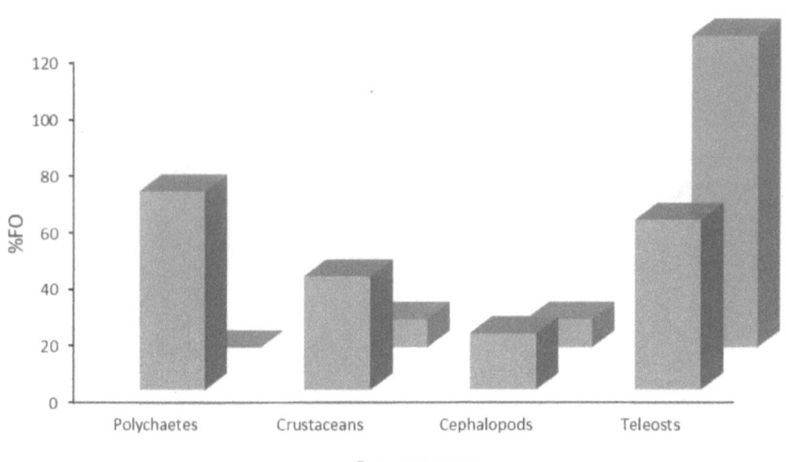

Figure 16: The percentage frequency of occurrence (*%FO*) of dietary categories in the diet of medium *Squalus acutipinnis* found at a 50 m depth bin on the West and South coasts of South Africa. The South coast diet data were subsampled and 10 stomachs were analysed on each coast.

Teleosts formed an important component in the diet of the medium sharks at the 150 m depth bin on the West coast, whilst crustaceans dominated the diet on the South coast (Fig. 17). For large sharks occurring at the 50 m depth bin on the West coast, polychaetes were important in their diet, whereas cephalopods were important on the South coast (Fig. 18). At the 150 m depth bin, however, teleosts were important in the diet of large sharks on the West coast, whilst polychaetes dominated the diet of these sharks on the South coast (Fig. 19). Evidently, the importance of prey categories in the diet of these sharks differed depending on the coast.

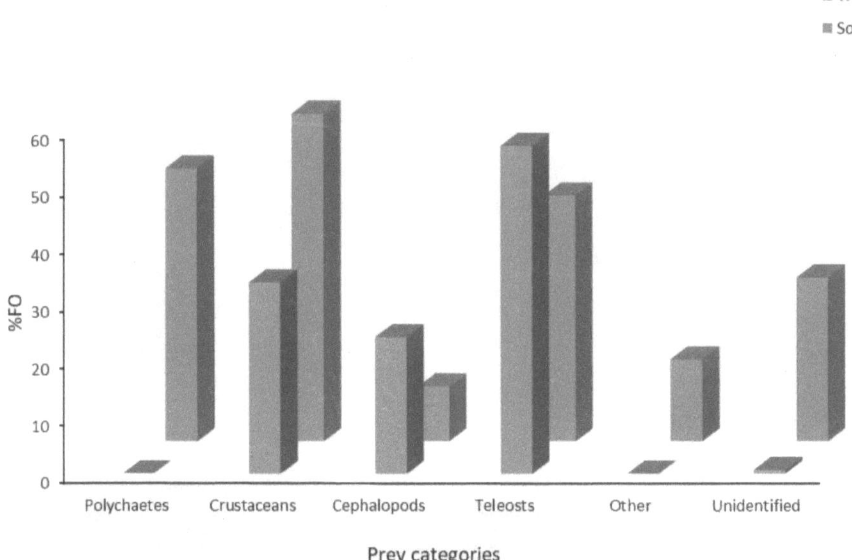

Figure 17: The percentage frequency of occurrence (*%FO*) of dietary categories in the diet of medium *Squalus acutipinnis* found at a 150 m depth bin on the West and South coasts of South Africa. The South coast diet data were subsampled and 21 stomachs were analysed on each coast.

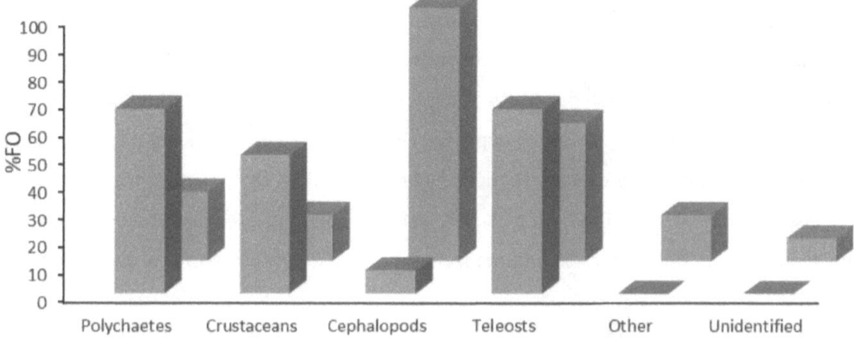

Figure 18: The percentage frequency of occurrence (*%FO*) of dietary categories in the diet of large *Squalus acutipinnis* found at a 50 m depth bin on the West and South coasts of South Africa. The South coast diet data were subsampled and 12 stomachs were analysed on each coast.

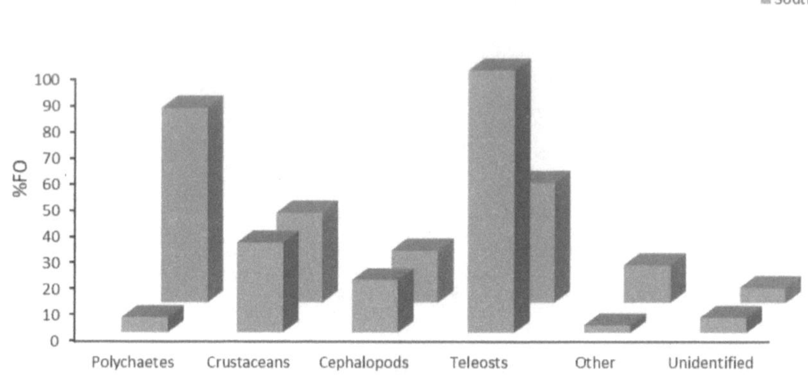

Figure 19: The percentage frequency of occurrence (*%FO*) of dietary categories in the diet of large *Squalus acutipinnis* found at a 150 m depth bin on the West and South coasts of South Africa. The South coast diet data were subsampled and 35 stomachs were analysed on each coast.

On the West coast, medium as well as large sharks caught in the 50 m, 150 m and 250 m depth bins, showed significant differences among the *%FO* of prey categories consumed (medium: $\chi^2 = 17.095$, *df* = 8; $p < 0.05$; large: $\chi^2 = 18.314$, *df* = 6, $p < 0.05$). Polychaetes dominated the diet of medium sharks sampled in the 50 m depth bin, whereas those sampled in the 150 m depth bin, fed on substantial amounts of both crustaceans and teleosts, whilst sharks collected in the 250 m depth bin predominantly consumed teleosts (Fig. 20). For large sharks, polychaetes dominated the diet at 50 m, teleosts at the 150 m depth bin and both cephalopods and teleosts in the 250 m depth bin (Fig. 21). Further significant differences in the *%FO* of prey categories of large sharks caught in the 150 m and 250 m depth bins only, were also found ($\chi^2 = 11.327$, *df* = 4, $p < 0.05$). Teleosts dominated the diet of these sharks at both depths but yet differed in the importance of alternative prey categories at each depth bin (Fig. 22).

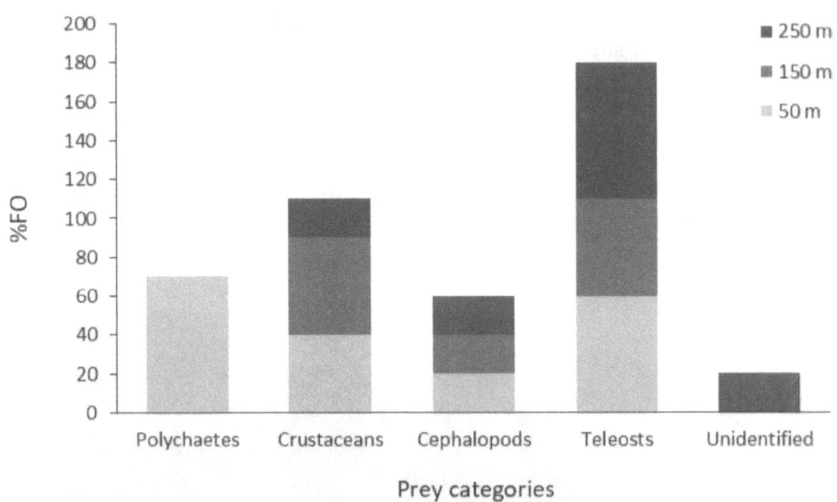

Figure 20: The percentage frequency of occurrence (*%FO*) of dietary categories in the diet of medium *Squalus acutipinnis* found at 50 m, 150 m and 250 m depth bins on the West Coast. The 150 m and 250 m diet data were subsampled and 10 stomachs were sampled for each depth bin.

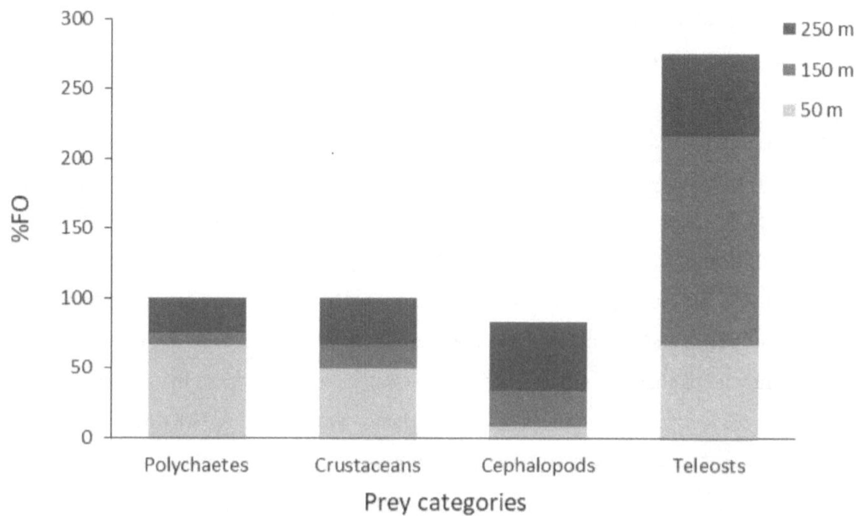

Figure 21: The percentage frequency of occurrence (*%FO*) of dietary categories in the diet of large *Squalus acutipinnis* found at 50 m, 150 m and 250 m depth bins on the West Coast. The 150 m and 250 m diet data were subsampled and 12 stomachs were analysed for each depth bin.

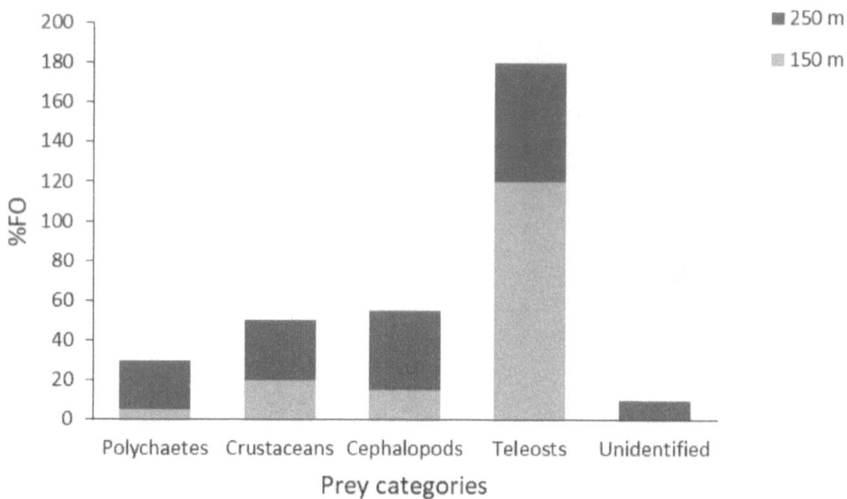

Figure 22: The percentage frequency of occurrence (*%FO*) of dietary categories in the diet of large *Squalus acutipinnis* found at 150 m and 250 m depth bins on the West Coast. The 150 m depth bin diet data were subsampled and 20 stomachs were analysed for each depth bin.

On the South coast, small, medium and large sharks caught in the 50 m and 150 m depth bins, showed significant differences in *%FO* of prey categories consumed (small: χ^2 = 37.296, *df* = 5, *p* <0.05; medium: χ^2 =15.335, *df* = 5, *p* <0.05; large: χ^2 = 12.044, *df* = 5, *p* <0.05). Small sharks fed predominantly on polychaetes and teleosts in the 50 m and 150 m depth bins, respectively (Fig. 23). Medium sharks, on the other hand, fed predominantly on teleosts in the 50 m depth bin and on both cephalopods and teleosts in the 150 m depth bin (Fig. 24), while teleosts, alone, contributed substantially to the diet of large sharks at both depths (Fig. 25). Moreover, significant differences were further evident among large sharks caught in the 150 m and 250 m depth bins (χ^2= 8.182, *df* = 5, *p* >0.05). Here, teleosts dominate the diet of sharks sampled at 150 m and polychaetes dominate the diet at 250 m (Fig. 26). Across tests, the overall diets of these sharks were comprised of various dietary categories and showed differences in the consumption of these alternative categories.

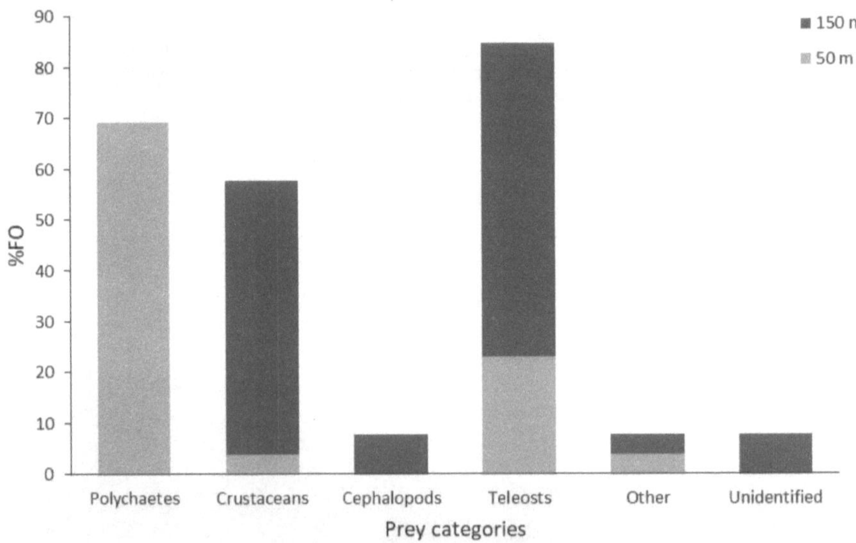

Figure 23: The percentage frequency of occurrence (*%FO*) of dietary categories in the diet of small *Squalus acutipinnis* found at 50 m and 150 m depth bins on the South Coast. The 150 m depth bin diet data were subsampled and 26 stomachs were analysed for each depth bin.

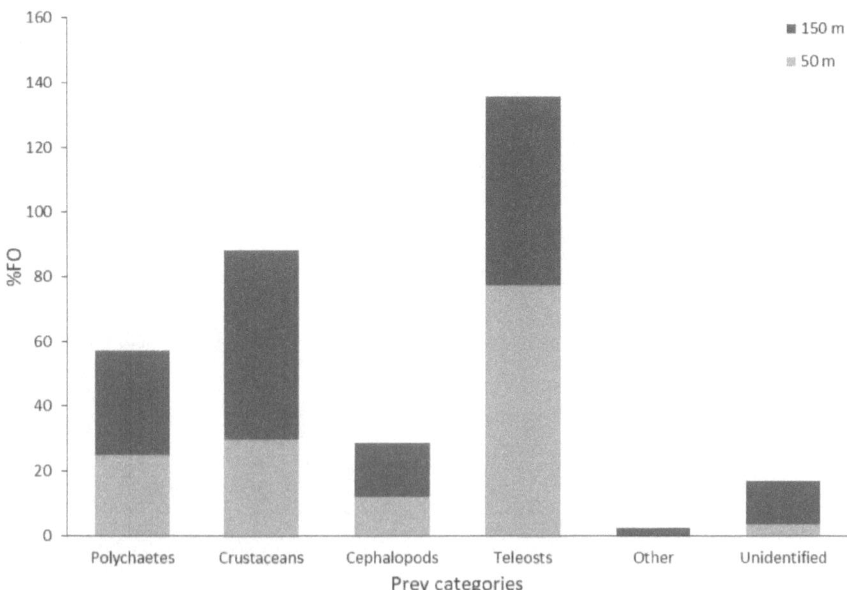

Figure 24: The percentage frequency of occurrence (*%FO*) of dietary categories in the diet of medium *Squalus acutipinnis* found at 50 m and 150 m depth bins on the South Coast. The 150 m diet data were subsampled and 84 stomachs were analysed for each depth bin.

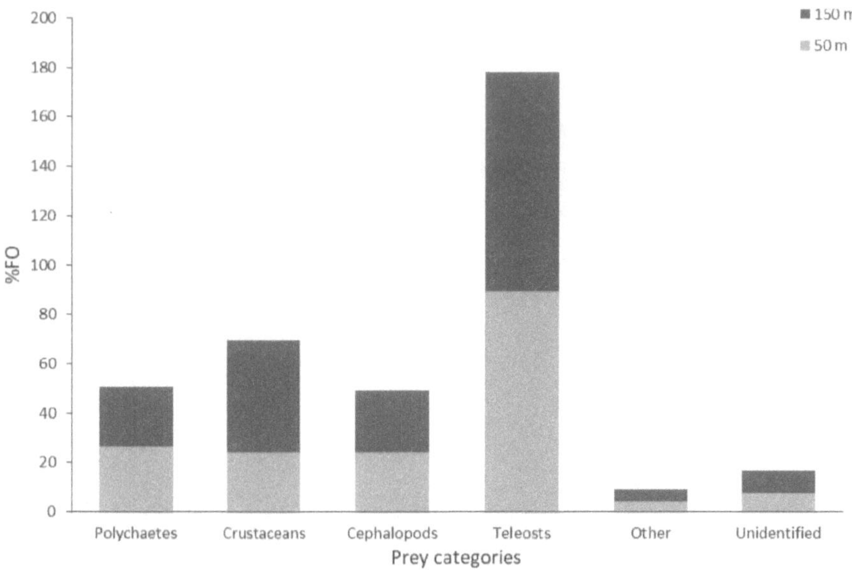

Figure 25: The percentage frequency of occurrence (*%FO*) of dietary categories in the diet of large *Squalus acutipinnis* found at 50 m and 150 m on the South Coast. The 150 m diet data were subsampled and 214 stomachs were analysed for each depth bin.

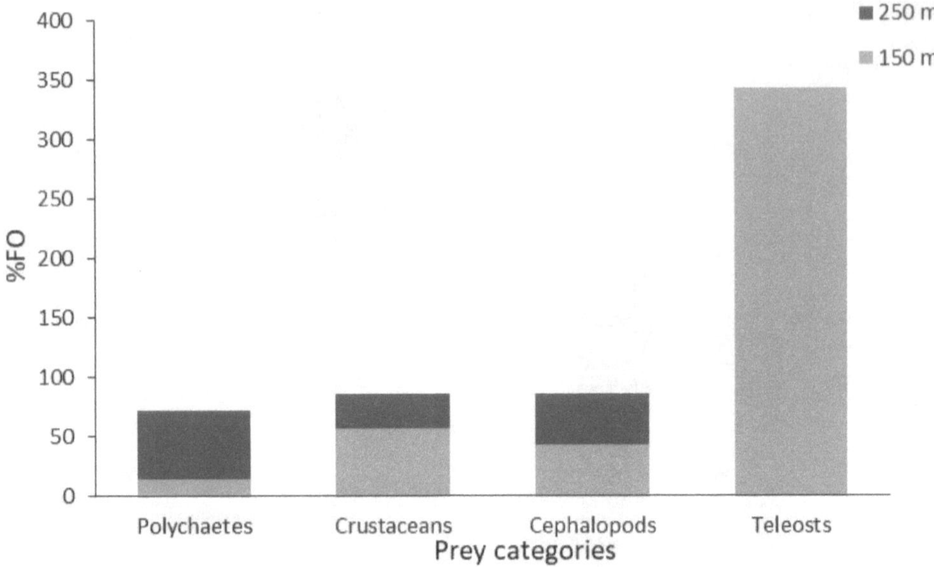

Figure 26: The percentage frequency of occurrence (*%FO*) of dietary categories in the diet of large *Squalus acutipinnis* found at 150 m and 250 m depth bins on the South Coast. The 150 m diet data were subsampled and 7 stomachs were analysed for each depth bin.

On the West coast, significant differences in the *%FO* of prey categories consumed by medium and large sharks were observed at the 250 m ($\chi^2 = 10.054$, $df = 4$, $p < 0.05$) depth bin only. Teleosts and cephalopods dominated the diet of medium and large sharks with differences stemming from the consumption of alternative categories (Appendix 6; Table 19).

Table 19: Dietary indices of number (*%N*), weight (*%W*), frequency of occurrence (*%FO*) and index of relative importance (*%IRI*) of large-sized *Squalus acutipinnis*, found at a 250 m depth bin on the West Coast of South Africa. This dataset was subsampled from the original and a total of 12 stomachs were analysed.

Phylum Annelida	**12.1**	**5.7**	**25.0**	**9.8**
Unidentified polychaete	12.1	5.7	25.0	9.8
Phylum Arthropoda	**19.0**	**9.7**	**30.0**	**7.0**
Funchalia woodwardi	1.7	3.0	5.0	0.5
Pterygosquilla armata capensis	6.9	6.3	10.0	2.9
Unidentified crustacean	10.3	0.4	15.0	3.6
Phylum Mollusca	**13.8**	**12.0**	**40.0**	**17.5**
Todaropsis eblanae	1.7	2.1	5.0	0.4
Unidentified cephalopod	12.1	9.9	35.0	17.0
Phylum Chordata: Osteichthyes	**51.7**	**63.6**	**60.0**	**62.9**
Callionymidae	1.7	2.6	5.0	0.5
Gnathophis sp.	1.7	20.6	5.0	2.5
Lampricoides	15.5	16.6	5.0	3.6
Unidentified teleost	32.8	23.8	45.0	56.4
Unidentified semi-digested material	**3.4**	**9.0**	**10.0**	**2.8**
Unidentified invertebrate	3.4	9.0	10.0	2.8
SUM	**100.0**	**100.0**	**165.0**	**100.0**

On the South coast, significant differences in the *%FO* of prey categories, among small, medium and large sharks were evident at both 50 m ($\chi^2 = 28.035$, *df* = 8, $p < 0.05$) and 150 m ($\chi^2 = 20.974$, *df* = 10, $p < 0.05$) depth bins. At the 50 m depth bin, unidentified polychaetes dominated the diet of small sharks (Appendix 2) whereas teleosts dominated the diet of medium and large sharks (Table 20 and 21). On the other hand, at the 150 m depth bin, teleosts and crustaceans dominated the diet of small sharks (Appendix 3) whereas teleosts dominated the diet of medium and large sharks, respectively (Table 22 and 23). In addition to this, significant differences in the *%FO* of prey categories among only medium and large sharks caught in the 50 m ($\chi^2 = 30,429$, *df* = 5, $p < 0.05$) and 150 m ($\chi^2 = 11,570$, *df* = 5, $p < 0.05$) depth bins were further evident. Although teleosts dominated the diet of both medium (Appendix 7) and large (Table 24) sharks at the 50 m depth bin, they differed in the consumption of alternative prey categories (Appendix 7; Table 24). Moreover, polychaetes and teleosts collectively dominated the diet of medium sharks (Appendix 8), while teleosts dominated the diet of large sharks at the 150 m depth bin (Table 25). The overall diets of these sharks were comprised of various dietary categories and showed differences in the consumption of these alternative categories.

Table 20: Dietary indices of number (*%N*), weight (*%W*), frequency of occurrence (*%FO*) and index of relative importance (*%IRI*) of medium-sized *Squalus acutipinnis*, found at 50 m on the South Coast of South Africa. This dataset was subsampled from the original and a total of 26 stomachs were analysed

Species	%N	%W	%FO	%IRI
Phylum Annelida	**22.92**	**7.25**	**26.92**	**25.56**
Unidentified polychaetes	22.92	7.25	26.92	25.56
Phylum Arthropoda	**8.33**	**1.00**	**15.39**	**1.13**
Decapoda	2.08	0.28	3.85	0.29
Goneplax angulata	2.08	0.28	3.85	0.29
Penaeidae	2.08	0.28	3.85	0.29
Unidentified crustacean	2.08	0.17	3.85	0.27
Phylum Mollusca	**6.25**	**23.78**	**11.54**	**3.63**
Loligo reynaudii	2.08	2.79	3.85	0.59
Unidentified cephalopod	2.08	0.36	3.85	0.30
Unidentified octopus	2.08	20.63	3.85	2.75
Phylum Chordata: Osteichthyes	**62.50**	**67.97**	**84.62**	**69.68**
Austroglossus pectoralis	4.17	4.46	3.85	1.04
Engraulis encrasicolus	14.58	20.54	19.23	21.25
Genypterus capensis	2.08	3.12	3.85	0.63
Merluccius sp.	4.17	1.12	3.85	0.64
Sardinops sagax	4.17	13.10	7.69	4.18
Trachurus capensis	8.33	13.63	11.54	7.97
Unidentified teleost	25.00	12.00	34.62	33.96
SUM	**100.00**	**100.00**	**138.46**	**100.00**

Table 21: Dietary indices of number (*%N*), weight (*%W*), frequency of occurrence (*%FO*) and index of relative importance (*%IRI*) of large-sized *Squalus acutipinnis*, found at 50 m on the South Coast of South Africa. This dataset was subsampled from the original and a total of 26 stomachs were analysed

Species	%N	%W	%FO	%IRI
Phylum Annelid	**36.21**	**5.60**	**34.62**	**49.45**
Unidentified polychaete	36.21	5.60	34.62	49.45
Phylum Arthropoda	**13.79**	**1.66**	**26.92**	**2.49**
Decapoda	1.72	0.21	3.85	0.25
Goneplax angulata	1.72	0.42	3.85	0.28
Mysid	3.45	0.08	3.85	0.46
Pterygosquilla armata capensis	1.72	0.63	3.85	0.31
Unidentified amphipod	1.72	0.25	3.85	0.26
Unidentified crustacean	3.45	0.06	7.69	0.92
Phylum Mollusca	**10.35**	**27.01**	**23.08**	**12.92**
Loligo reynaudii	5.17	20.62	11.54	10.17
Sepia australis	3.45	5.91	7.69	2.46
Unidentified cephalopod	1.72	0.48	3.85	0.29
Phylum Chordata: Osteichthyes	**39.66**	**65.74**	**76.92**	**35.13**
Austroglossus pectoralis	1.72	0.42	3.85	0.28
Clupeidae	1.72	0.84	3.85	0.34
Engraulis encrasicolus	12.07	3.89	15.39	8.39
Etrumeus whiteheadi	3.45	3.37	7.69	1.79
Merluccius sp.	1.72	1.05	3.85	0.37
Paracallionymus costatus	1.72	0.32	3.85	0.27
Pterogymnus laniarius	1.72	1.68	3.85	0.45
Sardinops sagax	3.45	5.26	7.69	2.29
Trachurus capensis	1.72	39.55	3.85	5.43
Unidentified teleost	10.35	9.36	23.08	15.54
SUM	**100.00**	**100.00**	**161.54**	**100.00**

Table 22: Dietary indices of number (*%N*), weight (*%W*), frequency of occurrence (*%FO*) and index of relative importance (*%IRI*) of medium-sized *Squalus acutipinnis*, found at 150 m on the South Coast of South Africa. This dataset was subsampled from the original and a total of 35 stomachs were analysed

Species	%N	%W	%FO	%IRI
Phylum Annelid	**28.77**	**7.97**	**34.29**	**34.54**
Unidentified polychaete	28.77	7.97	34.29	34.54
Phylum Arthropoda	**9.59**	**2.57**	**17.14**	**1.22**
Decopoda	1.37	1.69	2.86	0.24
Goneplax angulata	2.74	0.63	5.71	0.53
Mysida	2.74	0.04	2.86	0.22
Unidentified crustacean	1.37	0.13	2.86	0.12
Unidentified isopod	1.37	0.08	2.86	0.11
Phylum Mollusca	**5.48**	**4.78**	**11.43**	**0.80**
Loligo reynaudii	1.37	2.11	2.86	0.27
Nassarius vinctus	1.37	0.84	2.86	0.17
Unidentified cephalopod	1.37	1.53	2.86	0.23
Unidentified mollusc	1.37	0.30	2.86	0.13
Phylum Chordata: Osteichthyes	**54.80**	**83.41**	**91.43**	**63.24**
Austroglossus pectoralis	2.74	3.38	2.86	0.48
Engraulis encrasicolus	16.44	20.73	28.57	29.12
Etrumeus whiteheadi	4.11	13.60	2.86	1.39
Macrura sp.	1.37	0.08	2.86	0.11
Merluccius capensis	1.37	0.84	2.86	0.17
Merluccius sp.	2.74	0.84	2.86	0.28
Paracallionymus costatus	1.37	0.84	2.86	0.17
Sardinops sagax	2.74	15.61	5.71	2.88
Trachurus capensis	6.85	14.53	11.43	6.70
Unidentified teleost	15.07	12.94	28.57	21.94
Unidentified semi-digested mate	**1.37**	**1.27**	**2.86**	**0.21**
Unidentified specimen	1.37	1.27	2.86	0.21
SUM	**100.00**	**100.00**	**157.14**	**100.00**

Table 23: Dietary indices of number (*%N*), weight (*%W*), frequency of occurrence (*%FO*) and index of relative importance (*%IRI*) of large-sized *Squalus acutipinnis*, found at 150 m on the South Coast of South Africa. This dataset was subsampled from the original and a total of 35 stomachs were analysed

Species	%N	%W	%FO	%IRI
Phylum Annelida	**25.58**	**10.51**	**17.14**	**29.10**
Unidentified polychaete	25.58	10.51	17.14	29.10
Phylum Arthropoda	**15.12**	**9.58**	**28.57**	**4.77**
Funchalia woodwardi	1.16	0.07	2.86	0.17
Goneplax angulata	3.49	0.49	8.57	1.60
Mysida	2.33	0.06	2.86	0.32
Pleisionika martia	1.16	0.28	2.86	0.19
Pterygosquilla armata capensis	2.33	0.49	5.71	0.76
Unidentified crustacean	1.16	0.03	2.86	0.16
Upogebia sp.	3.49	8.18	2.86	1.57
Phylum Mollusca	**8.14**	**19.73**	**17.14**	**6.64**
Loligo reynaudii	1.16	0.14	2.86	0.18
Sepia australis	1.16	0.28	2.86	0.19
Todaropsis eblanae	3.49	18.03	5.71	5.78
Unidentified cephalopod	1.16	0.67	2.86	0.25
Unidentified octopus	1.16	0.62	2.86	0.24
Phylum Chordata: Elasmobranchii	**1.16**	**2.08**	**2.86**	**0.44**
Rajidae	1.16	2.08	2.86	0.44
Phylum Chordata: Osteichthyes	**40.70**	**57.07**	**88.57**	**52.12**
Belonidae	1.16	3.05	2.86	0.57
Bregmaceros sp.	1.16	0.04	2.86	0.16
Chelidonichthys queketti	1.16	0.28	2.86	0.19
Clupeidae	1.16	0.56	2.86	0.23
Engraulis encrasicolus	11.63	4.51	20.00	15.18
Etrumeus whiteheadii	3.49	5.69	8.57	3.70
Gnathophis sp.	3.49	3.91	8.57	2.98
Merluccius sp.	1.16	9.71	2.86	1.46
Paracallionymus costatus	3.49	0.76	5.71	1.14
Scomber japonicus	1.16	22.88	2.86	3.23
Unidentified teleost	11.63	5.69	28.57	23.27
Unidentified semi-digested material	**9.30**	**1.03**	**14.29**	**6.94**
Unidentified specimen	9.30	1.03	14.29	6.94
SUM	**100.00**	**100.00**	**168.57**	**100.00**

Table 24: Dietary indices of number (*%N*), weight (*%W*), frequency of occurrence (*%FO*) and index of relative importance (*%IRI*) of large-sized *Squalus acutipinnis*, found at 50 m on the South Coast of South Africa. This dataset was subsampled from the original and a total of 84 stomachs were analysed

Species	%N	%W	%FO	%IRI
Phylum Annelida	**1.36**	**0.24**	**2.38**	**0.19**
Unidentified polychaete	1.36	0.24	2.38	0.19
Phylum Arthropoda	**12.93**	**0.78**	**21.43**	**1.66**
Brachyura	0.68	0.10	1.19	0.05
Caridea	0.68	0.01	1.19	0.04
Decapoda	2.04	0.17	3.57	0.39
Goneplax angulata	0.68	0.10	1.19	0.05
Mysida	2.72	0.07	3.57	0.50
Parapaguridae	1.36	0.10	2.38	0.17
Penaeidae	0.68	0.07	1.19	0.05
Plesionika martia	0.68	0.10	1.19	0.05
Pterygosquilla armata capensis	0.68	0.02	1.19	0.04
Unidentified amphipod	1.36	0.03	2.38	0.17
Unknown crustacean	1.36	0.01	2.38	0.16
Phylum Mollusca	**14.97**	**29.28**	**25.00**	**9.99**
Aplysiidae	0.68	0.07	1.19	0.05
Loligo reynaudii	5.44	8.87	9.52	6.79
Afrololigo mercatoris	0.68	0.31	1.19	0.06
Octopus vulgaris	1.36	4.35	2.38	0.68
Sepia australis	1.36	9.49	1.19	0.64
Sepia sp.	0.68	0.19	1.19	0.05
Todaropsis eblanae	0.68	0.24	1.19	0.05
Unidentified cephalopod	2.72	1.58	4.76	1.02
Unidentified octopus	1.36	4.19	2.38	0.66
Phylum Echinodermata	**0.68**	**0.14**	**1.19**	**0.05**
Holothuroidea	0.68	0.14	1.19	0.05
Phylum Chordata: Tunicata	**0.68**	**2.95**	**1.19**	**0.22**
Unidentified tunicate	0.68	2.95	1.19	0.22
Phylum Chordata: Chondrichthyes	**4.76**	**20.91**	**8.33**	**1.52**
Rajidae	0.68	0.71	1.19	0.08
Selachii	0.68	1.00	1.19	0.10
Squalus acutipinnis	0.68	0.33	1.19	0.06
Torpedo sinuspersici	0.68	13.28	1.19	0.83
Torpedo sp.	0.68	2.13	1.19	0.17
Unidentified chondrichthyan	0.68	0.29	1.19	0.06
Unidentified elasmobranch	0.68	3.18	1.19	0.23
Phylum Chordata: Osteichthyes	**58.50**	**43.36**	**75.00**	**83.65**
Belonidae	0.68	1.04	1.19	0.10
Bregmaceros sp.	0.68	0.10	1.19	0.05
Clupeidae	0.68	1.42	1.19	0.13
Cynoglossus zanzibarensis	0.68	0.26	1.19	0.06
Engraulis encrasicolus	24.49	8.73	20.24	33.48
Etrumeus whiteheadii	1.36	0.09	2.38	0.09
Gnathophis sp.	0.68	0.12	1.19	0.05
Helicolenus dactylopterus	0.68	7.05	1.19	0.46
Merluccius sp.	0.68	2.13	1.19	0.17
Paracallionymus costatus	4.08	0.38	4.76	1.06
Sardinops sagax	2.72	5.53	4.76	1.96
Trachurus capensis	2.72	3.92	3.57	1.18
Unidentified teleost	18.37	12.58	30.95	44.88
Unidentified semi-digested material	**6.12**	**1.55**	**7.14**	**2.73**
Unidentified specimen	6.12	1.55	7.14	2.73
SUM	**100.00**	**100.00**	**141.67**	**100.00**

Table 25: Dietary indices of number (*%N*), weight (*%W*), frequency of occurrence (*%FO*) and index of relative importance (*%IRI*) of large-sized *Squalus acutipinnis*, found at 150 m on the South Coast of South Africa. This dataset was subsampled from the original and a total of 159 stomachs were analysed

Species	%N	%W	%FO	%IRI
Phylum Annelida	**15.42**	**6.60**	**28.93**	**21.88**
Nereididae	0.41	0.22	1.26	0.03
Unidentified polychaetes	15.01	6.38	27.67	21.85
Phylum Cnidaria	**0.41**	**1.45**	**1.26**	**0.09**
Pennatulaceae	0.41	1.45	1.26	0.09
Phylum Arthropoda	**22.52**	**4.79**	**45.28**	**6.45**
Caridea	0.61	0.02	1.26	0.03
Decapoda	0.61	0.08	1.89	0.05
Euphausiaceae	0.61	0.02	1.89	0.04
Funchalia woodwardi	0.41	0.19	1.26	0.03
Goneplax angulata	0.61	0.43	1.89	0.07
Malacostraca	0.41	0.05	1.26	0.02
Mursia cristiata	0.81	0.41	2.52	0.11
Mysida	1.83	0.31	3.77	0.30
Parapaguridae	1.01	0.22	3.15	0.14
Passihaea sppI. 1	0.20	0.11	0.63	0.01
Penaeidae	0.41	0.03	1.26	0.02
Pleisionika martia	0.20	0.11	0.63	0.01
Pterygosquilla armata capensis	1.01	0.58	3.15	0.19
Unidentified amphipod	10.35	0.31	11.32	4.45
Unidentified crustacean	2.84	0.44	7.55	0.91
Upogebia africana	0.41	0.37	1.26	0.04
Upogebia sp.	0.20	1.11	0.63	0.03
Phylum Mollusca	**9.33**	**13.99**	**23.27**	**3.59**
Enteroctopus magnificus	0.20	0.51	0.63	0.02
Inioteuthis capensis	0.20	0.02	0.63	0.01
Loligo reynaudii	3.65	4.88	5.66	1.78
Sepia australis	0.20	0.95	0.63	0.03
Sepia sp.	0.81	0.33	2.52	0.11
Sepiolidae	0.20	0.06	0.63	0.01
Teuthida	1.01	1.83	3.15	0.33
Todaropsis eblanae	0.20	0.89	0.63	0.03
Unidentified cephalopod	2.23	2.35	6.92	1.17
Unidentified octopus	0.41	2.14	1.26	0.12
Unidentified mollusc	0.20	0.04	0.63	0.01
Phylum Echinodermata	**0.20**	**0.11**	**0.63**	**0.01**
Holothuroidea	0.20	0.11	0.63	0.01
Phylum Chordata: Chondrichthyes	**0.81**	**6.17**	**2.52**	**0.65**
Unidentified elasmobranch	0.81	6.17	2.52	0.65
Phylum Chordata: Osteichthyes	**51.32**	**66.90**	**96.86**	**67.34**
Austroglossus pectoralis	0.20	0.22	0.63	0.01
Clupeidae	0.20	0.45	0.63	0.02
Cynoglossus zanzibarensis	0.41	0.39	1.26	0.04
Engraulis encrasicolus.	5.07	3.22	11.95	3.66
Etrumeus whiteheadii	4.46	7.28	5.66	1.06
Helicolenus dactylopterus	0.20	0.11	0.63	0.01
Lampricoides	2.64	0.22	0.63	0.07
Lophius vomerinus	0.20	3.05	0.63	0.08
Merluccius capensis	0.41	1.23	1.26	0.08
Merluccius sp.	0.41	1.77	1.26	0.10
Paracallionymus costatus	12.37	2.76	9.43	5.27
Sardinops sagax	4.46	10.58	7.55	4.19
Scomber japonicus	0.20	0.59	0.63	0.02
Scomberesox saurus scomberoides	0.20	2.08	0.63	0.05
Trachurus capensis	1.01	8.38	3.15	1.09
Unidentified teleost	14.60	17.41	39.62	46.84
Unidentified semi-digested material	**4.26**	**7.16**	**11.32**	**4.77**
Unidentified specimen	4.26	7.16	11.32	4.77
SUM	**100.00**	**100.00**	**198.74**	**100.00**

Teleosts, cephalopods, crustaceans, and polychaetes were important prey categories in the diet of *Squalus acutipinnis*, with the degree of importance changing with coast, depth and size class. Overall, small-sized *S. acutipinnis* fed predominantly on unidentified polychaetes, with larger sharks focussing on cephalopods, crustaceans and most importantly, teleosts, with anchovy *Engraulis encrasicolis* being the most frequently consumed teleost. Although overlap in the prey items per dietary category was evident, the prey items' importance varied in number, weight and frequency of occurrence consumed. As the size of fish increased, there was an increase in the number of different prey taxa observed irrespective of coast, with larger sharks displaying broader diets than their smaller counterparts.

Intraspecific evaluation of resource-use by Squalus bassi

A total of 67 different prey taxa were found in the stomachs of *Squalus bassi*: 29 on the West coast and 61 on the South coast. Teleosts were the most important prey category in the overall diet of these sharks, with cephalopods and crustaceans of lesser importance (Table 26). The diets of males and females were pooled, and no analyses by sex were made as no differences in the diet by sex was found (p<0.001) when inferences could be made. Moreover, sex-related differnces in diet could also not be tested across all tests because of insufficient sample numbers.

Table 26: The number of different prey items by prey category found in the overall diet of *Squalus bassi* on the West and South Coast of South Africa

Prey Categories	West coast	South coast
Polychaetes	0	3
Cephalopods	6	14
Teleosts	16	28
Crustaceans	4	13
Other	2	1
Unidentified semi-digested material	1	2

Medium *Squalus bassi* were only caught in the 250 m (N= 6; Appendix 16) depth bin on the West coast, whereas these sharks were caught in the 150 m (N= 24; Appendix 17); 250 m (N= 6; Appendix 18) and 450 m (N= 6; Appendix 19) depth bins on the South coast. Based on *%IRI*, teleosts and cephalopods were the most important prey categories for sharks caught at all depth bins, except for the 250 m depth bin on the South coast, where crustaceans replaced cephalopods in terms of dietary importance (Appendices 16-19).

Large *Squalus bassi* were caught at 150 m (N= 14; Appendix 20); 250 m (N= 32; Appendix 21); 350 m (N= 14; Appendix 22) and 550 m (N= 12; Appendix 23) on the West coast and at 150 m (N= 57; Appendix 24); 250 m (N= 31; Appendix 25); 450 m (N= 87; Appendix 26); 550 m (N= 55; Appendix 27) and 650 m (N= 6; Appendix 28) on the South coast. Based on the *%IRI* and in order of greatest importance, teleosts and cephalopods were the most important prey categories in the diet of sharks collected at all depth bins (Appendix 20 – 27), except the 650 m depth bin on the South coast, where cephalopods were more important than teleosts (Appendix 28).

There were no significant differences in the *%FO* of prey categories, between the West and South coasts for medium sharks occurring at 250 m ($\chi^2 = 3.058$, $df = 2$, $p > 0.05$), or between large sharks occurring at 150 m ($\chi^2 = 3.735$, $df = 3$, $p > 0.05$) and 250 m ($\chi^2 = 0.625$, $df = 4$, $p > 0.05$) depth bins. When results were not significant, details are not presented.

On the West coast, large sharks showed no significant differences in the *%FO* of prey categories caught in the 150 m, 250 m and 350 m depth bins ($\chi^2 = 3.058$, $df = 2$, $p > 0.05$), or between those caught in the 150 m and 250 m depth bins, only ($\chi^2 = 1.586$, $df = 3$, $p > 0.05$).

On the South coast, however, comparisons among medium sharks caught in the 150 m, 250 m and 450 m depth bins, showed significant differences in the *%FO* of prey categories ($\chi^2 = 19.607$, $df = 10$, $p < 0.05$). Sharks caught in the 150 m depth bin, fed predominantly on crustaceans and cephalopods, while those collected at 250 m fed on substantial amounts of teleosts. In the 450 m depth bin, sharks fed mainly on cephalopods (Fig. 27).

Significant differences were also evident in the *%FO* of prey categories of large sharks caught in the 150 m, 250 m, 450 m and 550 m depth bins ($\chi^2 = 28.049$, $df = 15$, $p < 0.05$). Teleosts were the most important prey category in the 150 m, 250m and 550 m depth bins, while both teleosts and cephalopods were most important diet category of sharks sampled in the 450 m depth bin (Fig. 28). When testing for differences in the *%FO* of prey categories of large sharks caught between 150 m and 250 m depth bins only, no differences were evident ($\chi^2 = 7,493$ $df = 4$, $p > 0.05$); while significant differences were evident for sharks caught between 450 m and 550 m depth bins only ($\chi^2 = 9,981$, $df = 4$, $p < 0.05$). Here too, teleosts were the dominant prey category in the diet of sharks, with the absence of alternative categories possibly driving differences (Fig. 29). The overall diets of these sharks comprised a variety of categories and showed differences in the consumption of these alternative categories (Fig. 27 - 29).

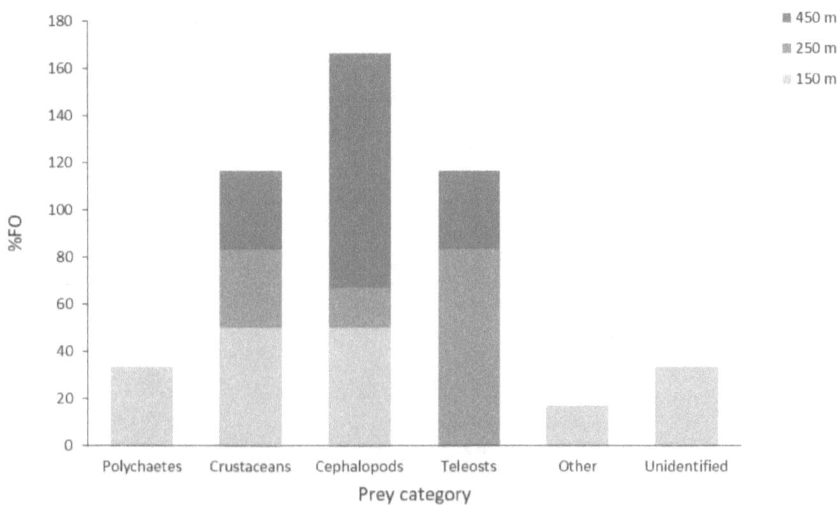

Figure 27: The percentage frequency of occurrence (*%FO*) of dietary categories in the diet of medium *Squalus bassi* found at 150 m, 250 m and 450 m depth bins on the South coast. The 150 m diet data were subsampled and a total of 6 stomachs were analysed per depth bin.

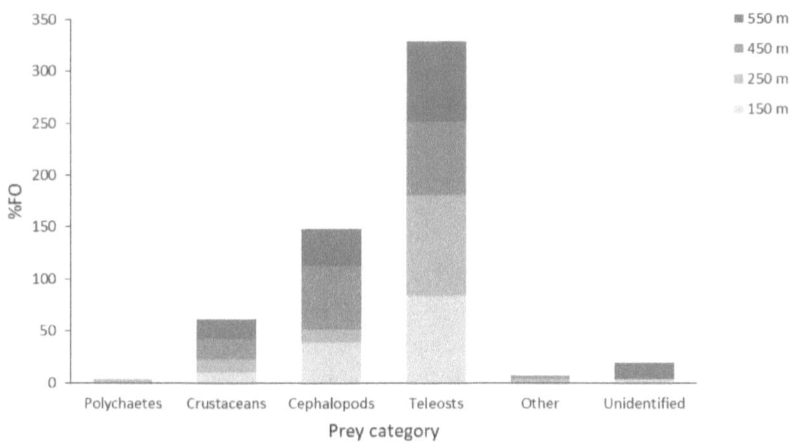

Figure 28: The frequency of occurrence (*%FO*) of dietary categories in the diet of large *Squalus bassi* found at 150 m, 250 m, 450 m and 550 m on the South coast. The diet data for 150 m, 450 m and 550 m were subsampled and a total of 31 stomachs were analysed.

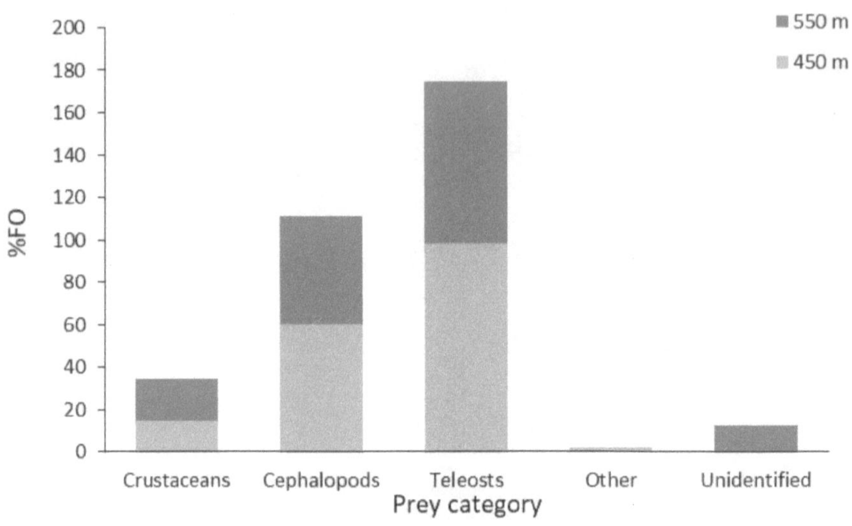

Figure 29: The percentage frequency of occurrence (*%FO*) of dietary categories in the diet of large *Squalus bassi* found at 450 m and 550 m depth bins on the South coast. The 450 m depth bin diet data were subsampled and a total of 55 stomachs were analysed.

There were no significant differences in the *%FO* of prey categories among medium and large sharks occurring in the 250 m depth bin on the West coast ($\chi^2 = 7.106$, $df = 3$, p >0.05). On the South coast however, significant differences in the *%FO* of prey categories were evident among medium and large sharks occurring in the 150 m ($\chi^2 = 14.944$, $df = 5$, p <0.05) and 450 m ($\chi^2 = 6.857$, $df = 2$, p <0.05) depth bins. At the 150 m depth bin, although teleosts dominated the diet of both medium (Appendix 17) and large sharks (Table 27), they differed in the consumption of alternative prey categories, with medium sharks even feeding on a prey category (crustaceans) that large sharks did not feed on (Appendix 17; Table 27). At the 450 m depth, although cephalopods dominated the diets of both medium (Appendix 19) and large sharks (Table 28), medium sharks also fed on crustaceans whereas the large sharks had not. No differences were evident between medium and large sharks occurring at 250 m ($\chi^2 = 1.279$, $df = 2$, p > 0.05).

Table 27: Dietary indices of number (*%N*), weight (*%W*), frequency of occurrence (*%FO*) and index of relative importance (*%IRI*) of large-sized *Squalus bassi*, found at 150 m on the South Coast of South Africa. This dataset was subsampled from the original and a total of 24 stomachs were analysed

Species	%N	%W	%FO	%IRI
Phylum Arthropoda	**3.28**	**0.04**	**8.33**	**2.67**
Decapoda	1.64	0.03	4.17	1.34
Unidentified crustacean	1.64	0.01	4.17	1.33
Phylum Mollusca	**34.43**	**39.30**	**54.17**	**43.08**
Crossia sp.	1.64	0.36	4.17	1.61
Histiotheuthis macrohista	1.64	1.01	4.17	2.13
Loligo reynaudii	13.12	18.22	4.17	25.19
Sepia sp.	3.28	10.56	8.33	5.56
Teuthidae	3.28	5.75	8.33	3.63
Unidentified cephalopod	8.20	2.68	20.83	1.75
Unidentified octopus	3.28	0.72	4.17	3.21
Phylum Chordata: Osteichthyes	**59.02**	**54.19**	**87.50**	**46.41**
Gnathopis sp.	4.92	5.79	8.33	4.31
Lampricoides	1.64	0.18	4.17	1.46
Myctopidae	9.84	5.61	4.17	12.42
Paracallionymus costatus	8.20	1.29	4.17	7.63
Sardinops sagax	4.92	8.34	8.33	5.33
Scomberesox saurus scomberoides	4.92	6.35	12.50	5.20
Trachurus capensis	3.28	2.01	8.33	2.13
Trichiurus lepturus	3.28	6.04	8.33	3.74
Unidentified teleost	18.03	18.58	29.17	4.21
Unidentified semi-digested material	**3.28**	**6.47**	**8.33**	**7.84**
Unidentified invertebrate	1.64	0.72	4.17	1.90
Unidentified specimen	1.64	5.75	4.17	5.94
SUM	**100.00**	**100.00**	**158.33**	**100.00**

Table 28: Dietary indices of number (*%N*), weight (*%W*), frequency of occurrence (*%FO*) and index of relative importance (*%IRI*) of large-sized *Squalus bassi*, found at 450 m on the South Coast of South Africa. This dataset was subsampled from the original and a total of 6 stomachs were analysed

Species	%N	%W	%FO	%IRI
Phylum Mollusca	**75.00**	**50.45**	**100.00**	**90.99**
Unidentified cephalopod	75.00	50.45	100.00	90.99
Phylum Osteichthyes	**25.00**	**49.55**	**33.33**	**9.01**
Engraulis encrasicolus	12.50	40.57	16.67	6.42
Paracallionymus costatus	12.50	8.98	16.67	2.60
SUM	**100.00**	**100.00**	**133.33**	**100.00**

Teleosts, cephalopods and crustaceans were important prey categories in the diet of *Squalus bassi*, with the degree of importance changing with coast, depth and size class. Overall, *S. bassi,* irrespective of size, fed more often on teleosts than any other prey category. Cape hakes, *Merluccius* spp. were the identifiable teleost that was most frequently consumed. Although overlap in the dietary categories were evident, the importance of prey items consumed varied in number, weight and frequency of occurrence, as well as species. As with *S. acutipinnis* the bigger the sample fish the greater the number of different prey items consumed irrespective of coast, with larger sharks displaying broader diets than their smaller counterparts.

Interspecific evaluation of resource-use between Squalus acutipinnis *and* S. bassi

Differences in the diet of the two species were compared using *%FO* in prey categories, but because of sample size constraints, it could only be tested explicitly among large sharks of each species, occurring in the 150 m and 250 m depth bins on both coasts. The results indicated that the diets differed significantly in the 150 m (West: χ^2 = 14.775, *df*= 3, *p* <0.05; South: χ^2 = 33.210, *df* = 5, *p* <0.05) and 250 m (West: χ^2 = 16.154, *df* = 5, *p* <0.05; South: χ^2 = 6.606, *df* = 2, *p* <0.05) depth bins on both coasts.

At the 150 m depth bin on the West coast, crustaceans were the most important prey category in the diet of *Squalus acutipinnis* whereas teleosts were more important in the diet of *S. bassi* (Fig. 30a). In the 250 m depth bin, however, although teleosts and cephalopods were the most important prey categories in the diet of both species, their diet differed in relative amounts as well as by the presence/absence of alternative prey categories (Fig. 30b). A similar result was observed in the 150 m depth bin on the South coast, where teleosts and cephalopods were generally the most important prey categories in the diet of both species, but there were differences in the relative amounts as well as in the presence/absence of alternative prey categories (Fig. 30c). Cephalopods were most important in the diet of *S. acutipinnis* while polychaetes were the most important prey category in the diet of *S. bassi* in the 250 m depth bin (Fig. 30d).

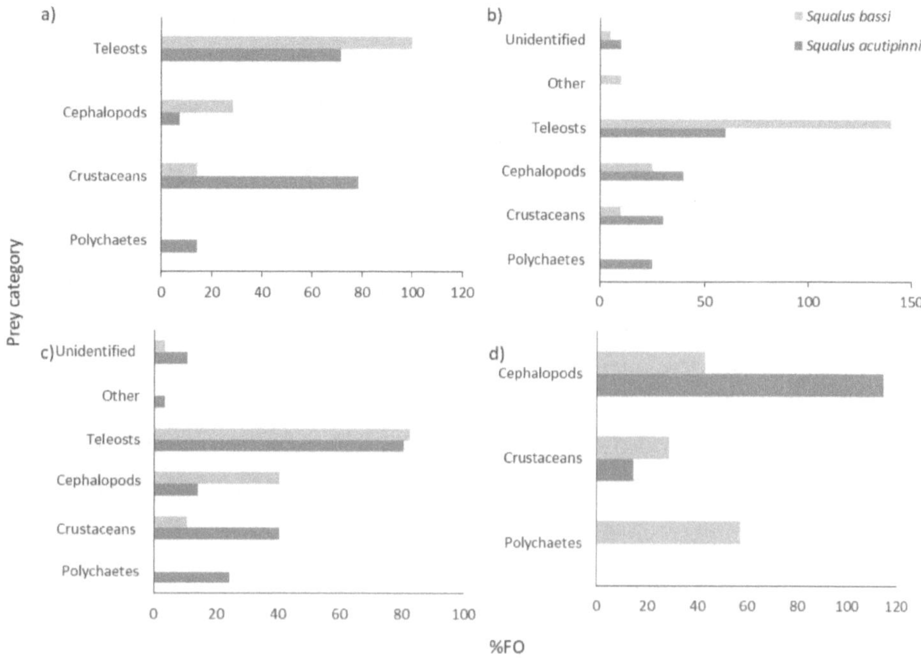

Figure 30: The percentage frequency of occurrence (*%FO*) in the dietary categories of large *Squalus acutipinnis* and large *Squalus bassi* found at the 150 m (a and c) and 250 m (b and d) depth bins on the West (a and b) and South (c and d) coasts of South Africa. Dietary data for the 150 m depth bins were subsampled for *S. acutipinnis*, while the dietary data at the 250 m depth bins were subsampled for *S. bassi*.

Discussion

This is the first comprehensive study on the feeding ecology of *Squalus acutipinnis* and *S. bassi*, along the West and South coasts of South Africa. The stomachs of many fish contained no food, which is not unusual (Cortés and Gruber, 1990; Compagno *et al.*, 1991; Kazunari, 1991; Ebert *et al.*, 1992; Bush, 2003; Braccini *et al.*, 2005; Varela *et al.*, 2013), as sharks are intermittent feeders that actively feed for only short periods of time (Wetherbee *et al.*, 1990, 2012; Wetherbee and Cortés, 2004; Braccini *et al.*, 2005). Regurgitation during capture could explain some of these results, as multiple sharks had water-filled stomachs once brought aboard the ship (Leslie RW, *pers comm.*, 2019, DFFE). The majority of stomachs that contained food, had few prey items, an observation in agreement with many others (Wetherbee *et al.*, 1990; Smale and Compagnon, 1997; Alonso *et al.*, 2001; Simpfendorfer *et al.*, 2001; Braccini *et al.*, 2005). This means that

81

a large number of stomachs are needed to describe diet with any confidence. Past studies have determined that a minimum of 200 stomachs would be needed to confidently describe the diet of a shark (Morato *et al.*, 2003; Braccini *et al.*, 2005): a number that was unfortunately not collected for every size class, depth class or coast here.

Squalus acutipinnis and *S. bassi* were shown to feed across all six prey categories and to consume a wide range of different prey species. They can, therefore, be classed as generalist feeders (Wetherbee *et al.*, 1990, 2012; Hanchet, 1991; Alonso *et al.*, 2002; Bethea *et al.*, 2004; Braccini *et al.*, 2005; Fanelli *et al.*, 2009; Huckembeck *et al.*, 2014). That said, teleosts were important prey for all size classes, becoming particularly important for larger fish. The importance of teleosts in the diet of sharks has been highlighted by many studies (Wetherbee *et al.*, 1990; Jones and Barmuta, 1998; Jakobsdóttir, 2001; Laptikhovsky *et al.*, 2001; Brickle *et al.*, 2003; Braccini *et al.*, 2005; Lucifora *et al.*, 2006; Dunn *et al.*, 2013; van der Heever, 2017). Of the identifiable teleost species, anchovy were important for *S. acutipinnis,* whereas hakes were important for *S. bassi*. It is noteworthy that anchovy were typically identified from heads found in the shark's stomach. Cephalopods usually behead fish when feeding, and the presence of fish heads may have resulted from the scavenging of cephalopod discards (as Smale and Compagno 1997). Although teleosts become more important with size, the opportunistic characteristic of these sharks is evidenced by the presence of small pelagic species (*Engraulis encrasicolus, Etrumeus whiteheadi, Sardinops sagax*) in the diet of large specimens of both species.

Some of the prey items noted here have not previously been recorded in the diet of the investigated species (Ebert *et al.*, 1992). This may reflect differences in the study site and/or sample size. The geographical area covered in this study ranges from the Orange River to Port Elizabeth, whereas that of Ebert *et al.* (1992) extended from Walvis Bay to Cape Agulhas. M*ursia cristiata* and *Engraulis encrasicolus* were recorded in the diet of *Squalus acutipinnis* on the West coast here, but not by Ebert *et al.* (1992). In the case of *S. bassi, Etrumeus whiteheadi* and *Symbolophorus* sp. were recorded in the diet of material collected on the West coast here, but not by Ebert *et al.* (1992). It should be realised, however, that the number of stomachs analysed in this study was much larger than that of Ebert *et al.* (1992), which would have allowed for a greater chance of encountering relatively uncommon, and hence newly recorded, prey species.

The lack of any relationships between the size of predator and prey is often inferred to be the result of opportunistic feeding by the predator and is commonly seen in studies of sharks (Wetherbee *et al.*, 1990). It has been suggested that sharks will consume (when hungry) any species of an appropriate size, that they encounter and they do not ignore one type of prey out of preference for another (Wetherbee *et al.*, 1990; Vögler *et al.*, 2009). Based on the data alone, a linear relationship between squaloid size and cephalopod size should not have been expected here, as the maximum size of many cephalopods species is much smaller (*Sepia* sp.) than the minimum size of squaloids. Moreover, squaloids have effective cutting teeth that work with a sawing action (Ebert, 2013; Ebert and Mostarda, 2013), therefore the size of prey, in general, may not be limited by gape size, but by the ability to capture the prey (Smale and Compagno, 1997). Even saying that these sharks are opportunistic feeders should be with some caution, because if the predator swallows its prey whole, the maximum size of prey that can be eaten will increase with increasing predator size whether the predator is an opportunistic feeder or not. The results clearly demonstrate that the diet of both species changes with increasing size, with small sharks switching from small (polychaetes) to bigger prey species (teleosts), meaning that effectively, the mean prey size, across all prey species, should increase with increasing predator size.

The results presented here indicate some significant differences in the diets of *Squalus acutipinnis* and *S. bassi* by coast, depth, and size. The differences in the diet of the same species between coasts likely reflect differences in prey diversity and availability between the two areas. Typically the number of different types of prey consumed on the South coast was greater than that consumed at the same depths on the West coast, a fact that can be attributed to the greater diversity of species in the former than latter regions (Branch and Branch, 2018). Regional differences in the diet of sharks have been commonly reported in the literature, and have been reported to reflect differences in prey field and availability as well as differing habitat utilisations between the regions (Cortés and Gruber, 1990; Yamaguchi and Taniuchi, 2000; Simpfendorfer *et al.*, 2001; Braccini *et al.*, 2005; Bethea *et al.*, 2006, 2007).

A study on the diet of *Mustelus palumbes* and *M. mustelus* around the coast of South Africa indicated a degree of regional differences in the diet of each species (Smale and Compagno, 1997). On the West coast, the diet of *M. palumbes* was dominated by hermit crabs, whereas stomatopods dominated the diet of these sharks on the South coast. On the other hand, *Jasus lalandii* were important in the diet of *M. mustelus* on the West coast while other prey were significant on the South coast. Smale and Compagno, (1997) attributed these differences to reflect regional differences in the distributions and availability of prey (Smale and Compagno, 1997). With further reference to the results of this study, regional differences also reflected differences in sample collection effort. More stomach samples were collected on the South coast, with the South coast diet data collected over 12 years, while the West coast data were only collected over a span of two years.

Intraspecific changes in diet by depth is commonly observed in squaloid sharks (Kazunari, 1991; Braccini *et al.*, 2005), though it is sometimes difficult to disentangle the effects of depth *per se*, from those associated with size. Deeper-living *Squalus acutipinnis* and *S. bassi* fed more on demersal fish (eg. *Merluccius* spp. and *Gnathophis* sp., etc.) and cephalopods than shallower living animals, though some pelagic fish were still present in the diet. Polychaetes decreased in dietary importance with increasing size and depth. Small and young *S. acutipinnis* are known to be bathypelagic (Compagno *et al.*, 1991), feeding in close connection with the seafloor (high number of annelids – polychaetes; crustaceans - *Funchalia woodwardi, Pterygosquilla armata capensis*) as well as in the upper water column (*Engraulis encrasicolus*; and other pelagic teleosts). Again, differences in the diet with depth can be attributed to prey availability and changes in the prey spectrum with changing depth, a conclusion also reached by Carrassón *et al. (*1992) studying the diet of *Galeus melastomus* in the western Mediterranean Sea.

But they can also be related to changes in size. Intraspecific size-based differences in the diet of sharks are well studied, with differences resulting from ontogenetic shifts (Wetherbee *et al.*, 1990; Kazunari, 1991; Lucifora *et al.*, 2000, 2006; Alonso *et al.*, 2002; Braccini *et al.*, 2005; Papastamatiou *et al.*, 2006; Bethea *et al.*, 2007; van der Heever, 2017). These ontogenetic shifts in the diet occur with changes in the dietary items consumed by sharks as they grow; i.e. small and large sharks exploit different resources. Overall, as *Squalus acutipinnis* grows, there is a change in commonly consumed prey items. Polychaetes are important for small sharks, while teleosts become significantly more important in the diet of the larger sharks. Besides the obvious, major change in prey categories with growth, discrete changes in teleost prey species consumed are also evident for medium (e.g. *Engraulis encrasicolus, Etrumeus whiteheadi*) and large (*Cynoglossus zanzibarensis, Trachurus capensis, Trichiuris lepturus*) sharks at 150 m on the West coast.

Similarly, the diets of small, medium and large *Squalus acutipinnis*, caught at 50 m on the South coast, show differences in the teleost species predated. The larger the shark, the more benthic teleosts were consumed (eg: *Austroglossus pectoralis*, *Genypterus capensis*, *Merluccius* spp. and *Trachurus capensis*), whereas smaller sharks consumed more pelagic prey. Differences in the teleost types consumed by *S. bassi* at 150 m on the South coast were also noted. Small sharks consumed *Engraulis encrasicolus*, while the larger sharks consumed more mesopelagic and demersal species (e.g., Myctophidae, *Trichiurus lepturus* and *Gnathopis* sp.). Differences in the prey types of secondarily important categories (crustaceans and cephalopods) are also recognised.

Differences in the diet of predators with increasing size may reflect changes in their gape size, speed and handling ability. The consumption of polychaetes by small sharks may be as a direct result of gape-limitations; animals with small teeth and gape sizes may consume slow-moving prey or by scavenging. As the shark grows it is increasingly able to consume bigger prey, reflecting an increase in gape size, increased swimming speed and better handling abilities (Lucifora *et al.*, 2000, 2006; Graham, 2005; Bethea *et al.*, 2006, 2007; Braccini, 2008; Barbini *et al.*, 2011). And this would then result in increased diversity in the fish consumed. In the case of *Squalus megalops* in southern and eastern Australia, smaller sharks fed on benthic crustaceans and gastropods, whereas larger animals fed on demersal-pelagic prey: a fact that was attributed, in part, to changes in foraging ability (Braccini *et al.*, 2005). In the case of *Galeocerdo cuvier* off Hawaii, small sharks had a less diverse diet than large sharks and there was a shift from fish to more mammals and turtles (Lowe *et al.*, 1996). Small sharks were even shown to segregate spatially from their medium and large conspecifics (Lowe *et al.*, 1996). The data reported in Chapter Two showed that similar patterns in distribution were evident of the West and South coast of South Africa. It should be remembered that although increased gape size may increase their effectiveness as predators, squaloids are not limited by this as they have quite effective cutting teeth, with some even attacking and feeding on prey in packs (Ebert, 2013; Ebert and Mostarda, 2013; Leslie RW, *pers comm.*, 2019, DFFE).

Segregation by sex is a prevalent characteristic of shark life histories (Springer, 1967) and squaliform sharks are generally shown to distribute by sex (Kazunari and Tanaka, 1988; Kazunari, 1991; Braccini *et al.*, 2005; Ebert, 2013). However, the absence of sex-related differences in the diet of *Squalus* observed here has been reported before (Braccini *et al.*, 2005). These latter authors reasoned this to reflect small sample sizes (Ferry and Cailliet, 1996; Braccini *et al.*, 2005), which may also hold for this study.

The results indicate the lack of a clear interspecific overlap in resource-use by coast, depth, and size between the two subject species here, which could be interpreted as a result of interspecific competition (Schoener, 1974; Bethea *et al.*, 2004; Espinoza *et al.*, 2012; O'Shea, 2012; van der Heever, 2017). These species may be dividing their dietary resources, to enable co-existence through differential resource-use between coasts and depths, ontogenetic shifts between the size classes and the near-specialised resource-use by size classes (Schoener, 1974; Garrison and Link, 2000; Linke *et al.*, 2001; Platell and Potter, 2001; Yick *et al.*, 2011; Espinoza *et al.*, 2012, 2015). However, the term "competition" should be used with caution as the results do not indicate that resources are limited within the environment (Schoener, 1974; Ross, 1986).

Coexistence may be achieved if the two species consume the same resources (categories/prey species types), but in different proportions, or if one species consumes a prey category that the other species does not. Co-existence may be also be mediated if the two species use (even) slightly different micro-habitats (White and Potter, 2004; Marshall *et al.*, 2008). When species cooccur differences are pronounced and by consuming prey in different proportions, they may be facilitating co-existence. Off the east coast of North America, two similar-sized skates, *Raja erinacea* and *R. ocellata* were found to consume the same prey species but in varying proportions (McEachran *et al.*, 1976). The diet of *R. erinacea* was composed largely of teleosts and polychaetes whereas the diet of *R. ocellata* was largely composed of crustaceans. The difference in their diet became more pronounced when they co-occurred, with *R. erinacea* feeding on more epifaunal prey than *R. ocellata,* which fed on more infaunal species (McEachran *et al.*, 1976). However, based on the catch records, these species do seem to also occupy different depths and therefore different prey groups may become more or less available at the different depths. This difference may therefore also relate to habitat separations, rather than competition.

The results of this study have shown the intraspecific and interspecific dietary habits of *Squalus acutipinnis* and *S. bassi*. Their generalist, and perhaps opportunistic diets, complicates accurate description as South Africa moves towards ecosystem-based fishery management. It should be realised that using stomach content analyses as the sole method of quantifying the feeding ecology of species, does come with various limitations and is more often than not quite biased (Estrada *et al.*, 2006; Michener and Kaufman, 2007). This sampling technique is lethal, time-consuming and prey identification can be difficult, depending on the level of digestion. It also requires many stomachs to be analysed but yet only provides a snapshot of the last meal consumed, providing no real temporal information (Michener and Kaufman, 2007; Shiffman *et al.*, 2014; Espinoza *et al.*, 2015), which is complemented here by the use of stable isotopes analyses, which will be covered in Chapter Four.

Chapter Four

Intraspecific and interspecific variability in the isotopic signatures (δ^{13}C and δ^{15}N), of *Squalus acutipinnis* and *S. bassi*, caught off the South and West coast of South Africa

Introduction

Traditionally, information on the feeding ecology of sharks has been assessed through stomach content analyses (SCA; White *et al.*, 2004; Braccini, 2008; Vaudo and Heithaus, 2011; Baker *et al.*, 2014; Churchill *et al.*, 2015; Bonnici *et al.*, 2018; Chapter Three). This method involves identifying prey items directly from the stomach, allowing a high resolution of prey identification. Stomach content analysis is a lethal process and only provides a snapshot into the diet of sharks and is biased by when the animal last fed (Shiffman *et al.*, 2012, 2014; Valls *et al.*, 2017; Chapter Three).

Stomach content analyses have a number of other limitations (Estrada *et al.*, 2006; Michener and Kaufman, 2007). They are time-consuming to perform as they require specimen dissection and the identification of prey species (Estrada *et al.*, 2006; Michener and Kaufman, 2007). As sharks rarely consume the prey whole, often chunks of unidentifiable flesh are found in the stomachs, as evident in Chapter Three. Prey species are digested at different rates, and stomach content analyses are biased towards prey with hard bodies, resulting in the over-estimation of these prey in the importance of the diet, over the soft-bodied and easily digested species. In addition to this, stomach content analyses disregard assimilated prey (Estrada *et al.*, 2006; Michener and Kaufman, 2007). To overcome the natural variability in prey species consumed, stomach fullness and feeding plasticity of sharks, a large number of stomachs needs to be analysed before obtaining a comprehensive diet (Estrada *et al.*, 2006; Michener and Kaufman, 2007). This becomes particularly important when conducting a spatiotemporal study, with the inclusion of ontogenetic effects, but enough stomachs are rarely collected, with many even being empty (Braccini *et al.*, 2005; Dicken *et al.*, 2017) On the other hand, despite its importance, information on the trophic ecology is rarely provided by stomach content analyses (Cortés, 1999), further limiting the understanding of the full ecological role sharks have, and it provides no temporal information (MacNeil, *et al.*, 2005; Schmidt *et al.*, 2007; Valls *et al.*, 2017).

Stable isotope analysis (SIA) provides an alternative, complementary approach to understanding the diet and feeding ecology of species. SIA has a long history of use by terrestrial ecologists, anthropologists and physiologists, as a tool for detecting the sources of contaminants in natural systems, assessing the physiological condition of animals, tracking large-scale species migrations, determining food web structures as well as determining environmental temperatures and life-history characteristics of organisms of the past (Gannes *et al.*, 1997; Michener and Kaufman, 2007). Of late, SIA has also proven to be a useful technique for analysing these same processes within the marine environment (Fisk *et al.*, 2002; Domi *et al.*, 2005; Malpica-Cruz *et al.*, 2012; Valls *et al.*, 2017).

The use of SIA to better characterise marine food web dynamics has been of particular interest to marine ecologists and fishery scientists (Shiffman *et al.*, 2012; Hussey *et al.*, 2014; Churchill *et al.*, 2015). And sharks, as a component of marine ecosystems, have been included in some studies (Domi *et al.*, 2005; Logan and Lutcavage, 2010; Hussey *et al.*, 2011; Churchill *et al.*, 2015). Most of the above-listed studies have used the stable isotopes of carbon (δ^{13}C, ratio of ^{13}C/^{12}C, expressed relative to a standard) and nitrogen (δ^{15}N, the ratio of ^{15}N/^{14}N, expressed relative to a standard) as the tool to describe the food-web linkages within the environment, as these isotopes help determine carbon and nitriogen transfer between prey and their predators, and hence their relative trophic position (DeNiro and Epstein, 1978; Post, 2002; Pethybridge, 2010).

In summary, this methodology relies on the predictable fractionation process (enrichment of heavy isotopes) of the naturally occurring isotopes of carbon and nitrogen, effectively resulting in the predator having a higher proportion of heavy:light isotopes than that of the prey it consumes. Thus, the isotopic composition of a predator's tissue reflects that of its prey (DeNiro and Epstein, 1978; Minagawa and Wada, 1984; Domi *et al.*, 2005; Peterson, 2014). The stable isotope of carbon (δ^{13}C) has proved to fractionate at a low, but constant rate, increasing at <1 ‰ from prey to predator, per trophic position, and δ^{13}C is typically used to infer the source of primary productivity (i.e., terrestrial or marine; nearshore or offshore; benthic or pelagic). By contrast, fractionation of the stable isotope of nitrogen (δ^{15}N) has been shown to increase by 3‰ - 4 ‰, from prey to predator, per trophic position, and it is used to determine the relative trophic position of a species (DeNiro and Epstein, 1978; Peterson and Fry, 1987; Pethybridge, 2010; Matich, 2014).

While SIA may not be able to provide the detailed resolution about the diet that stomach content analysis can, the isotopic composition of tissue reflects assimilation (Michener and Kaufman, 2007; Logan and Lutcavage, 2010). Importantly, it represents an integration of diet over time, as tissue turnover rates range from days (blood) to months (white muscle) and even years (vertebrae), which allows for the monitoring of the long-term feeding behaviour of species (Logan and Lutcavage, 2010; Matich *et al.*, 2011; Hussey *et al.*, 2012a; Matich and Heithaus, 2014). Moreover, $\delta^{13}C$ signatures can be used to distinguish feeding location, i.e. inshore vs offshore feeding (Estrada *et al.*, 2003). Other advantages of SIA include the fact that animals need not be killed to get samples (only a small piece of tissue is required), and that every sample collected can provide data – there is no need for a full stomach.

Squalus acutipinnis and *S. bassi*, are common and co-occurring mesopredatory sharks, and based off trawl diversity make up the second most significant proportion of the demersal fauna of southern African waters (Ebert *et al.*, 1992; Watson and Smale, 1999; Richardson *et al.*, 2000; Veríssimo *et al.*, 2011). Despite this, comprehensive information on their feeding and trophic ecology is largely lacking. The results of Chapter Three indicated that *S. acutipinnis* and *S. bassi* fed significantly on teleosts, crustaceans, cephalopods and polychaetes, with the degree of importance of each prey taxon varying intraspecifically and interspecifically, by coast, size and depth. Differences in diet by sex were not recorded. Significant differences in the diets of these sharks with ontogeny mirrored depth occupation and teleosts became important with size. Recorded changes in the diet by coast likely reflected the change in the ambient food environment and even though there were marked intraspecific and interspecific differences, it was difficult to attribute to competition alone. Although a total of 1 433 stomachs were dissected for stomach content analysis (Chapter Three), this number was not enough to comprehensively describe the diet of both species.

Therefore, to fully understand the ecological role and impact that *Squalus acutipinnis* and *S. bassi* may have on their ecosystems, comprehensive knowledge of their trophic ecology is required. This study aims to elucidate the trophic ecology of *S. acutipinnis* and *S. bassi* using stable isotope analyses. Specifically, it aims to examine changes in the isotopic signatures of $\delta^{15}N$ and $\delta^{13}C$ concerning coast (location), sex, depth and size classes, from both an intraspecific and interspecific perspective. It was hypothesized that:

H_0: There is no intraspecific variability in the isotopic signatures of $\delta^{15}N$ or $\delta^{13}C$ by coast, sex, depth and size for each species.

H_0: There is no interspecific variability in the isotopic signatures of $\delta^{15}N$ or $\delta^{13}C$ by, coast, sex, depth and size

Materials and methods

Study survey area and sampling design

Dorsal muscle tissue was removed from specimens collected between 2014 – 2016 during
routine demersal hake biomass surveys conducted by DFFE on the West and South coasts of
South Africa (Fig. 1). An explanation of the study area as well as survey design follows
Payne *et al.* (1985) and is described in more detail in Chapter Two.

Sample preparation and laboratory analyses

Once the trawl net was retrieved, sharks were identified, sexed, measured and weighed,
following the protocols outlined in Chapter Two. Tissue located in front of the first dorsal fin
was excised, placed into glass scintillation vials and frozen at $-20°C$, else whole specimens
were blast frozen at -20°C for later processing in the laboratory. To avoid contamination,
scalpels were rinsed with deionized water (dH_2O) between samples.

In the laboratory, muscle tissue samples were thawed, the epidermis was removed and
the muscle tissue was washed with dH_2O and placed into a clean scintillation vial. The
presence of lipids in the tissues of sharks has been shown to negatively skew $\delta^{13}C$ values
(Post *et al.*, 2007; Schiffman *et al.*, 2012), and so lipids were first removed from samples
following Kim and Koch, (2012). Ten millilitres petroleum ether was pipetted into each vial,
which was then sonicated for 15 mins, after which the petroleum ether was decanted,
replaced and the sample sonicated once again for 15 mins: following decantation of the
petroleum ether, samples were sonicated in 10 ml dH_2O for a further 15 mins, after which the
dH_2O water was also decanted. This process was repeated three times for each sample.
Thereafter, tissue samples were freeze-dried for 72 hours, homogenized using a mortar and
pestle and transferred to tinfoil wrapped vials, prior to analysis.

All stable isotope analyses were conducted at either the Stable Light Isotope Unit
based at the University of Cape Town or the Stable Isotope Laboratory, Mammal Research

Institute based at the University of Pretoria. At both laboratories, aliquots of between 0.5 -
0.7 mg of freeze-dried muscle tissue were weighed into tin capsules that had been pre-
cleaned in toluene. The isotopic analysis was done on a Flash EA 1112 Series Organic
Elemental Analyser coupled to a Delta V Plus stable light isotope ratio mass spectrometer
(IRMS) via a ConFlo IV gas-controlled system (all equipment supplied by Thermo Fischer,
Bremen, Germany). Carbon and nitrogen values were expressed in delta notation (δ)
following the equation:

$$\delta X = [(R_{sample}/R_{standard}) - 1] \times 1000 \qquad\qquad \text{Eq 7}$$

where δX indicates the isotopic signature ($\delta^{13}C$ or $\delta\ ^{15}N$) in parts per thousand (‰),
while R indicates the respective ratios of heavy: light isotopes ($^{15}N/^{14}N$ or $^{13}C/^{12}C$) in the
sample and standard. A laboratory running standard (Merck Gel: $\delta^{15}N$=6.8‰, $\delta^{13}C$ = -
20.57‰, C%=43.83, N%=14.64) and blank sample were run after every 12 unknown
samples. Sample values were compared using "in-house standards." Carbon isotope values
were referenced relative to Vienna Pee-Dee Belemnite, while nitrogen isotope values were
referenced relative to atmospheric nitrogen. Analytical precision was calculated by
averaging the greatest standard error values across all carbon and nitrogen runs,
respectively (Newton, *pers. comm.*, 2016). Analytical precision was <0.09‰ for $\delta^{15}N$ and
<0.08‰ for $\delta^{13}C$.

When tissues had a C:N ratio > 3.5, even after chemical correction, $\delta^{13}C$ values were
corrected using the lipid-normalized correction equation of Post *et al. (*2007):

$$\delta^{13}C_{treated} = (\delta^{13}C_{untreated} - (3.32 + (0.99*C:N))) \qquad\qquad \text{Eq 8}$$

This equation is usually used on samples that had not been chemically treated.
However, even after chemical lipid extraction procedures, some of the samples still had C:N
values > 3.5. Since no C:N was greater than 6.5, it was possible to apply equation 8 (Miller,
pers comm., 2019; van der Lingen, *pers comm.*, 2019). Although the extraction of lipids can
alter $\delta^{15}N$ values, Hoffman and Sutton, (2010) have determined that this is not always so.

Data analyses

Owing to the uneven spread of samples across the depth and size classes used in Chapter Two and Chapter Three, these data have been further aggregated as either on-shelf (≤ 200 m) or off-shelf (>200 m), and as either small (≤ 40 cm) or large (> 40 cm), respectively.

Inferences on intraspecific and interspecific differences in the isotopic signatures of $\delta^{15}N$ and $\delta^{13}C$ were made using either parametric (General Linear Model, GLM), or non-parametric (Kruskal Wallis) statistical tests, following normality testing. The isotopic signatures of $\delta^{15}N$ and $\delta^{13}C$ were used as the dependent scale variables, while coast, depth and size classes were used as the independent categorical (fixed factor) variables. When depth and size were correlated, size class was only included as a covariate.

Bi-plots of the raw isotopic values of $\delta^{15}N$ and $\delta^{13}C$ were constructed, as too were boxplots which depicted the isotopic range of $\delta^{15}N$ and $\delta^{13}C$ by coast, depth and size class. The relationships between TL and $\delta^{15}N$ and $\delta^{13}C$ were separately investigated by linear regression analyses, by species and coast.

Normality in the data sets was tested using the Shapiro Wilks test. All statistical tests were performed in IBM SPSS 22, and statistical significance was deemed at an alpha value of 0.05. The effect of sex was omitted from all statistical analyses of diet, as no sex differences in the diet were recorded (chapter 3).

Results

A total of 196 individuals were sampled for SIA (Table 29), with more individuals of both species being sampled off the South coast (Table 29). *Squalus acutipinnis* were sampled at shallower depths on the South coast than on the West coast, whereas the opposite was true for *S. bassi* (Table 29). There were no statistical differences in the size (TL) of sharks sampled between the two coasts, for either species (*S. acutipinnis*: $F_{1,91} = 2.360$, $p > 0.05$; *S. bassi*: $H' = 0.001$, $p > 0.05$; Table 30). Sharks sampled on-shelf were, on average, smaller than those sampled off-shelf, with large sharks being more frequently sampled. This was true for both species on both coasts (Table 30). Moreover, small sharks sampled on the West coast were, on average, larger than those sampled on the South coast, for both species, while the opposite was true for the large sharks on each coast (Table 30).

Table 29: The overall sample sizes (N), mean shark sizes (cm; ±SE) and size ranges (cm; TL) of *Squalus acutipinnis* and *S. bassi* sampled, on the West and South coast of South Africa. Depth bins (m) from which sharks were sampled are also included and all sharks sampled were processed for stable isotope analyses in this study.

	S. acutipinnis		*S. bassi*	
	South coast	West coast	South coast	West coast
N	68	25	75	28
mean	46.8	50.8	65.9	66.2
±SE	1.60	2.34	2.47	3.46
size range (cm)	22.5 - 78	28 - 77	22 - 100	28 - 99
depth bin (m)	42 - 226	121 - 252	103 - 661	194 - 556

Table 30: The sample sizes (N), mean (cm; ±SE) shark sizes (TL; range also shown), and depth bins (m) from which samples were collected by depth ($\leq$ 200 m or > 200 m; upper); and the sample sizes (N), mean (cm; ±SE) fish sizes (TL; range also shown), and depth bins (m) from which samples were collected by size (small or large; lower), for *Squalus acutipinnis* and *S. bassi* sampled, on the West and South coast of South Africa.

		S. acutipinnis				*S. bassi*			
		South coast		West coast		South coast		West coast	
		$\leq$ 200 m	> 200 m	$\leq$ 200 m	> 200 m	$\leq$ 200 m	> 200 m	$\leq$ 200 m	> 200 m
Depth	N	55	13	8	17	20	55	6	2
	mean	44.83	54.92	45.13	53.41	48.85	71.94	58.17	
	±SE	1.719	3.453	2.754	3.029	4.452	2.504	8.424	
	range (cm)	22.5 - 74	40 - 78	28 - 52	39 - 77	22 - 80	39 - 100	28 - 74	
	range (m)	42- 200	209 - 226	121 - 196	204 - 252	103 - 182	222 - 661	194 - 196	2
		small	large	small	large	small	large	small	
Size	N	22	46	2	23	11	64	3	
	mean	32.46	53.60	33.50	52.26	30.91	71.77	34.00	
	±SE	0.957	1.496	5.5	2.26	1.509	2.115	3.055	
	range (cm)	22.5 - 39	40 - 78	28 - 39	40 - 77	22 - 39	40 - 100	28 - 38	
	range (m)	44 - 123	42 - 226	121 - 243	177 - 252	103 - 418	172 - 661	196 - 204	1

Intraspecific evaluation of $\delta^{15}N$ and $\delta^{13}C$ by Squalus acutipinnis

The isotopic biplots of $\delta^{15}N$ and $\delta^{13}C$ for *Squalus acutipinnis* are shown in Figure 31, indicating clustering of South coast samples to the right and West coast samples to the left; with some overlap. No obvious separation by size or depth is apparent. As the dataset for $\delta^{15}N$ was normally distributed, it was possible to run a GLM, which demonstrates significant differences by coast, depth and size (TL). Interestingly, when excluding the intercept, size contributed the most to the variance in δ^{15}N, followed by depth and then coast (Table 31).

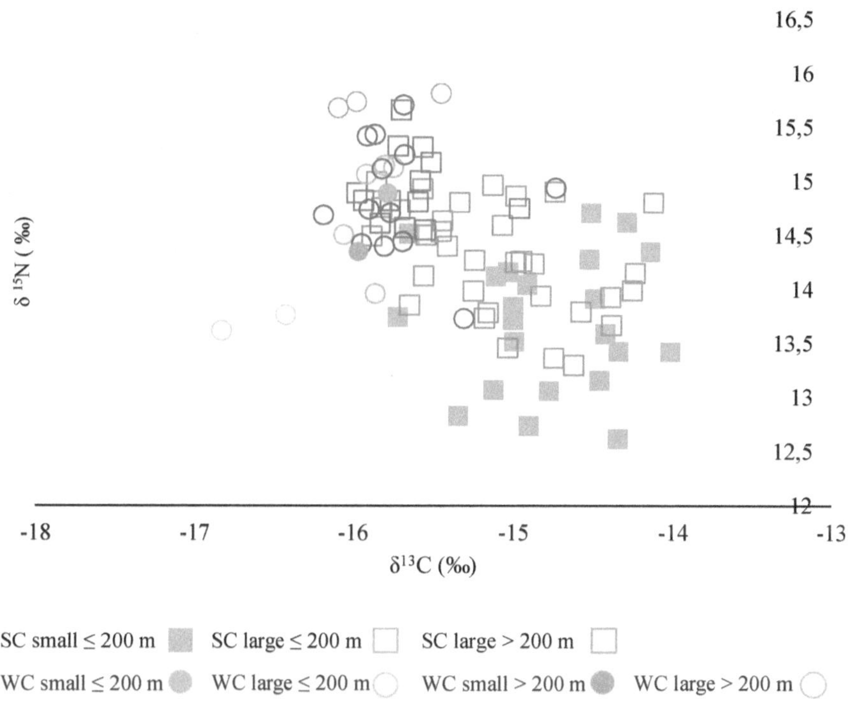

Figure 31: The isotopic signatures, expressed by delta notation in parts per thousand (‰), of ^{15}N (δ^{15}N) and ^{13}C (δ^{13}C), for small and large *Squalus acutipinnis* sharks, sampled at depths of ≤ 200 m (green) and > 200 m (blue) on the South (SC; square) and West coast (WC; circle) of South Africa.

Table 31: The results of a General Linear Model (GLM), which tested for significant differences in the variance of Nitrogen 15 (δ^{15}N‰), of *Squalus acutipinnis* sharks between coasts, depth classes and size (total length; TL). Significance is deemed at $p < 0.05$ and the level of significance is indicated by the use of an asterisk (i.e. $p <<0.05$ *; $p <<0.01$ **; $p <0.001$ ***)

Source	Sum of Squares	df	Mean Square	F	Sig.
Model	25.16	4	6.29	25.63	***
Intercept	900.05	1	900.05	3666.58	***
size (TL)	8.63	1	8.63	35.15	***
coast	0.98	1	0.98	3.98	*
depth	2.69	1	2.69	10.97	**
Error	21.60	88	0.25		

Sharks sampled on the West coast had, on average, higher $\delta^{15}N$ and lower $\delta^{13}C$ values than on the South coast (Figure 32a and 32b; Appendix 29). On the South and West coast, the mean $\delta^{15}N$ was greater at the depths > 200 m than at depths ≤ 200 m (Figure 33; Appendix 29). The results of the Kruskal Wallis test revealed no significant changes in $\delta^{13}C$ with depth on the West coast, but values were significantly lower for animals sampled at depths > 200 m than ≤ 200 m along the South coast (Figure 34; Appendix 29). Values of $\delta^{15}N$ were lower for small than large sharks on both coasts, and positive correlations were shown with size (Figs. 35 and 36). There were no size-related changes in $\delta^{13}C$ along the West coast, but decreases were seen with size on the South coast (Figs. 37 and 38).

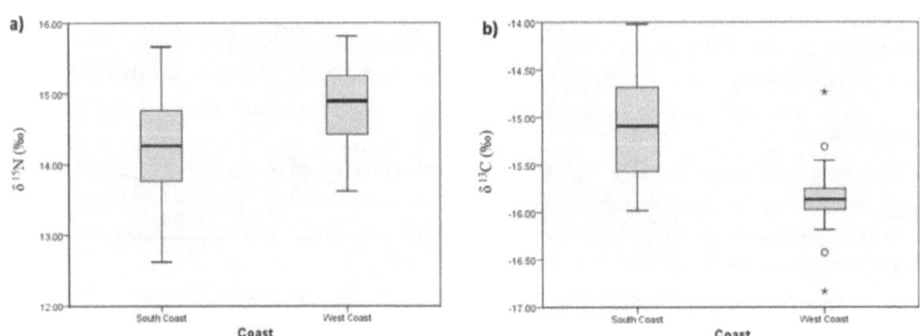

Figure 32: The range of isotopic signatures, expressed by delta notation in parts per thousand (‰), of ^{15}N ($\delta^{15}N$; **a)** and ^{13}C ($\delta^{13}C$; **b**), for *Squalus acutipinnis* sharks sampled on the South and West coast of South Africa. Significant differences were deemed at $p < 0.05$, and were evident between the coasts for $\delta^{15}N$ ($F_1 = 3.98$, $df = 1$; $p < 0.05$) and $\delta^{13}C$ ($H' = 33.21$; $df = 1$; $p < 0.001$). The boxplot provides a five number summary of the dataset, and depicts the minimum and maximum values of the dataset (lower and uppermost lines), the 25th and 75th percentile (bottom and top most inner lines) and the median (horizontal line through the center of the box). Outliers are represented by an asterisks (*) and circles(°).

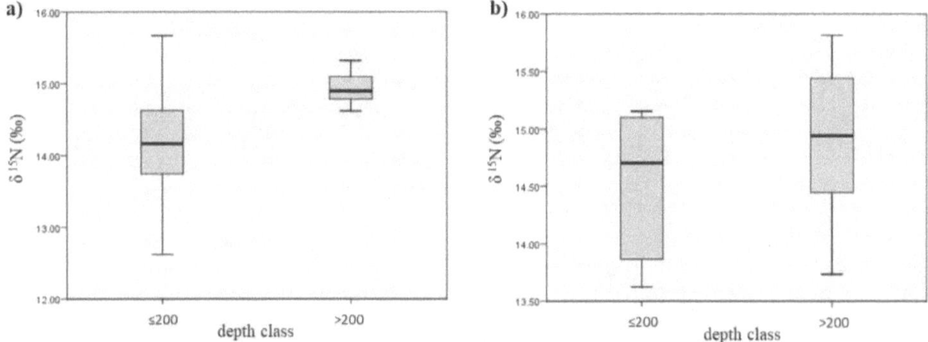

Figure 33: The range of isotopic signatures, expressed by delta notation in parts per thousand (‰), of ^{15}N (δ^{15}N), for *Squalus acutipinnis* sharks sampled at depths of ≤ 200 m and > 200 m on the South (**a**) and West coasts (**b**) of South Africa. Significant differences were deemed at p <0.05, and were evident between the depths for δ^{15}N (F_1 = 10.97, df = 1; p <0.01). The boxplot provides a five number summary of the dataset, and depicts the minimum and maximum values of the dataset (lower and uppermost lines), the 25th and 75th percentile (bottom and top most inner lines) and the median (horizontal line through the center of the box).

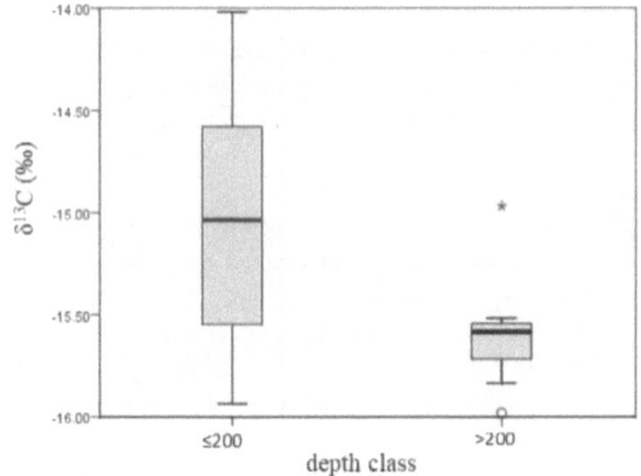

Figure 34: The range of isotopic signatures, expressed by delta notation in parts per thousand (‰) of ^{13}C (δ^{13}C), for *Squalus acutipinnis* sharks sampled at depths of ≤ 200 m and > 200 m on the South coast of South Africa. Significant differences were deemed at p <0.05, and were evident between depths on this coast (δ^{13}C: H' = 17.47; df = 1; p <0.001). The boxplot provides a five number summary of the dataset, and depicts the minimum and maximum values of the dataset (lower and uppermost lines), the 25th and 75th percentile (bottom and top most inner lines) and the median (horizontal line through the center of the box). Outliers are represented by an asterisk (*).

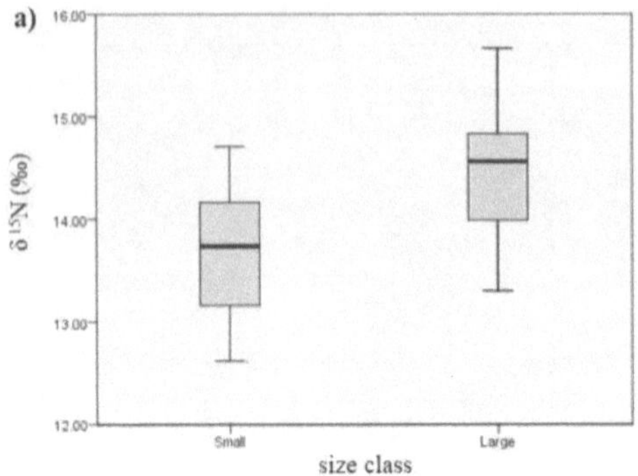 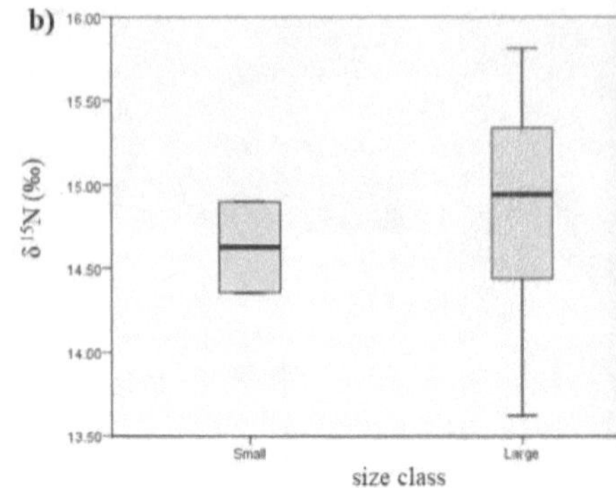

Figure 35: The range of isotopic signatures, expressed by delta notation in parts per thousand (‰), of ^{15}N (δ^{15}N), for small and large *Squalus acutipinnis* sharks sampled on the South (**a**) and West coast (**b**) of South Africa. Significant differences were deemed at $p < 0.05$, and were evident between the sizes for δ^{15}N ($F_1 = 3.98$, $df = 1$; $p < 0.0001$). The boxplot provides a five number summary of the dataset, and depicts the minimum and maximum values of the dataset (lower and uppermost lines), the 25th and 75th percentile (bottom and top most inner lines) and the median (horizontal line through the center of the box).

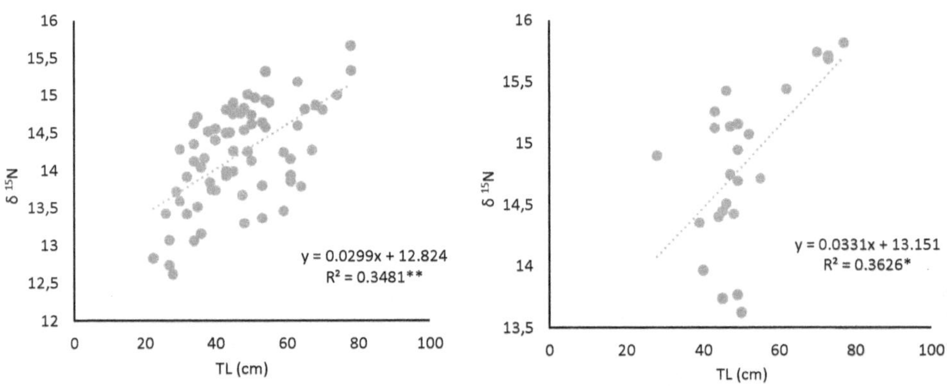

Figure 36: The scatterplot depicting the isotopic signature, expressed by delta notation in parts per thousand (‰), of ^{15}N (δ ^{15}N) vs size (TL; cm), for *Squalus acutipinnis* sharks sampled on the South (**a**) and West coast (**b**). Both graphs include an associated linear trendline equation and the R^2 value. Significance is deemed at $p < 0.05$ and the level of significance of the correlation is indicated by the use of an asterisk (i.e. $p < 0.05$ *; $p < 0.01$ **; $p < 0.001$ ***).

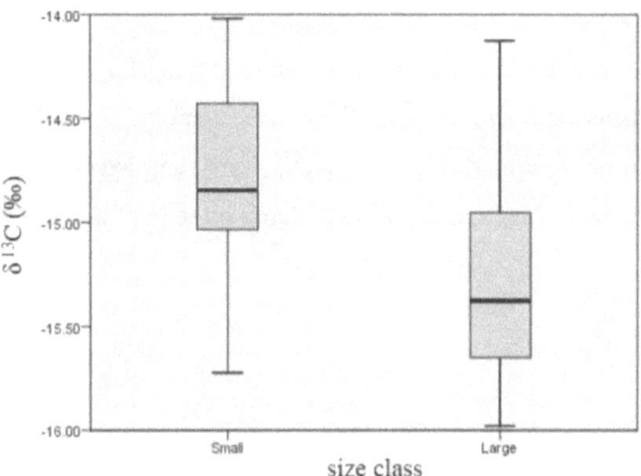

Figure 37: The range of isotopic signatures, expressed by delta notation in parts per thousand (‰), of ^{13}C (δ^{13}C), for small and large *Squalus acutipinnis* sharks sampled on the South coast of South Africa. Significant differences were deemed at $p < 0.05$, and were evident between the sizes for δ^{13}C ($H' = 10.66$; $df = 1$; $p < 0.001$). The boxplot provides a five number summary of the dataset, and depicts the minimum and maximum values of the dataset (lower and uppermost lines), the 25th and 75th percentile (bottom and top most inner lines) and the median (horizontal line through the center of the box).

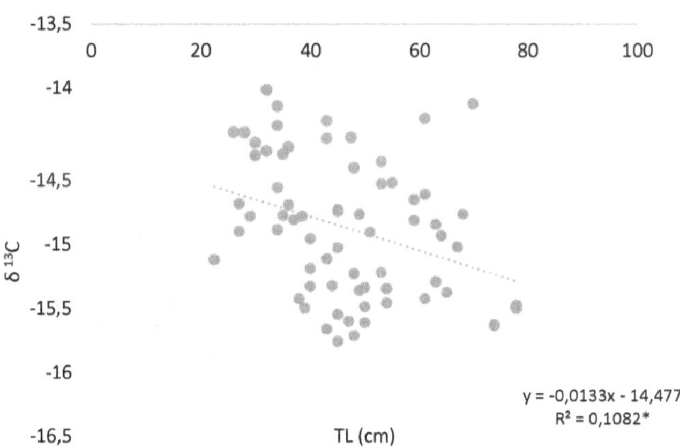

Figure 38: The scatterplot depicting the isotopic signature, expressed by delta notation in parts per thousand (‰), of ^{13}C (δ ^{13}C) vs size (TL; cm), for *Squalus acutipinnis* sharks sampled on the South coast of South Africa. The graph includes the associated linear trendline equation and the R^2 value. Significance is deemed at $p < 0.05$ and the level of significance of the correlation is indicated by the use of an asterisk (i.e. $p < 0.05$ *; $p < 0.01$ **; $p < 0.001$ ***).

***Intraspecific evaluation of $\delta^{15}N$ and $\delta^{13}C$ by* Squalus bassi**

The dataset collected for *Squalus bassi* was patchier than that of *S. acutipinnis* (Fig. 39). A certain amount of clustering is displayed by West coast samples, to the left, whereas South coast samples seem more scattered (Fig. 39). Small shark sampled at depths ≤ 200 m were separated from large sharks sampled at the same depth on the South coast (Fig. 39).

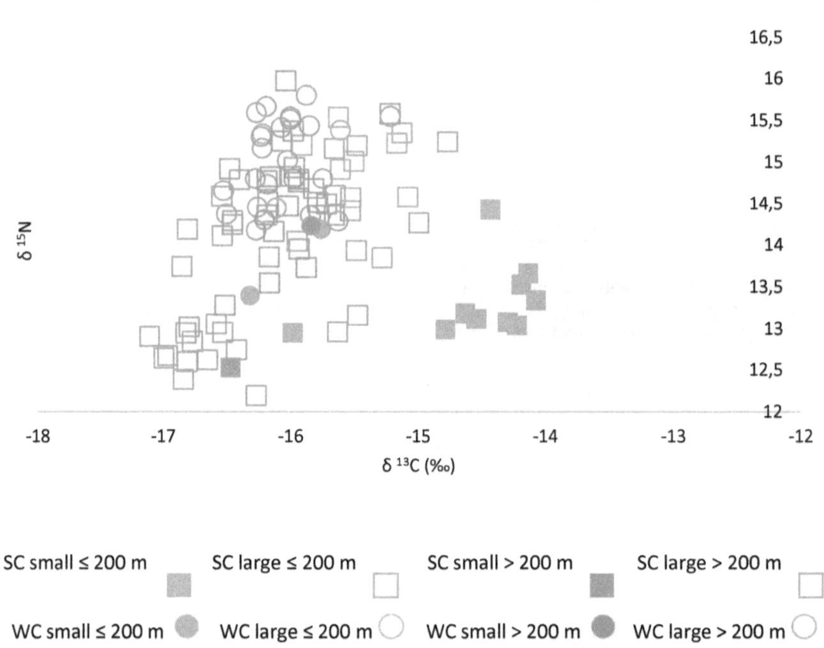

Figure 39: The isotopic signatures, expressed by delta notation in parts per thousand (‰), of [15]N (δ ^{15}N) and [13]C (δ^{13}C), for small (filled) and large (empty) *Squalus bassi* sharks, sampled at depths of ≤ 200 m (green) and > 200 m (blue) on the West (WC; circle) and South (SC; square) Coast of South Africa.

No other obvious separation is apparent. As the $\delta^{13}C$ dataset was normally distributed, it was possible to run a GLM, by coast, depth class and size (TL). When excluding the intercept, depth contributed the most to the variance in $\delta^{13}C$, followed by the interaction between coast and depth, and coast on its own (Table 32).

Table 32: The results of a General Linear Model (GLM), which tested for significant differences in the variance of Carbon 13 (δ^{13}C‰), of *Squalus bassi* sharks between coasts, depth classes and size (TL). Significance is deemed at $p <0.05$ and the level of significance is indicated by the use of an asterisk (i.e. $p <0.05$ *; $p <0.01$ **; $p <0.001$ ***; $p <0.0001$ ****)

Source	Sum of Squares	*df*	Mean Square	F	Sig.
Model	0.73	4	0.18	15.75	****
Intercept	0.54	1	0.54	46.61	****
size (TL)	0.04	1	0.04	3.08	>0.05
coast	0.06	1	0.064	5.57	*
depth	0.22	1	0.23	19.86	****
coast*depth	0.16	1	0.16	13.49	****
Error	1.13	98	0.01		

Sharks sampled on the West coast had, on average, higher δ^{15}N and lower δ^{13}C values than the sharks sampled on the South coast (Fig. 40a and 40b; Appendix 1). The results of the Kruskal Wallis tests revealed no change in δ^{15}N with depth on both coasts whereas significant changes in δ^{13}C were evident. Values of δ^{13}C were lower at depths > 200 m than at depths $\leq$ 200 m (Fig. 41a and 41b; Appendix 29). Significant changes in δ^{13}C values were further evident between sharks sampled at the same depth between the different coasts ($F_{1,98}$ = 13.49, $p <0.001$; Table 32). Sharks sampled at $\leq$ 200 m on the West coast had a lower, average δ^{13}C value, than those sampled on the South coast, while the reverse pattern was evident at depths > 200 m (Appendix 29). Large sharks had significantly higher δ^{15}N values than small sharks on both coasts, and positive correlations were shown with size (Fig. 42a, 42b, 43a and 43b; Appendix 29). There were no significant changes in δ^{13}C with size on both coasts (Table 32).

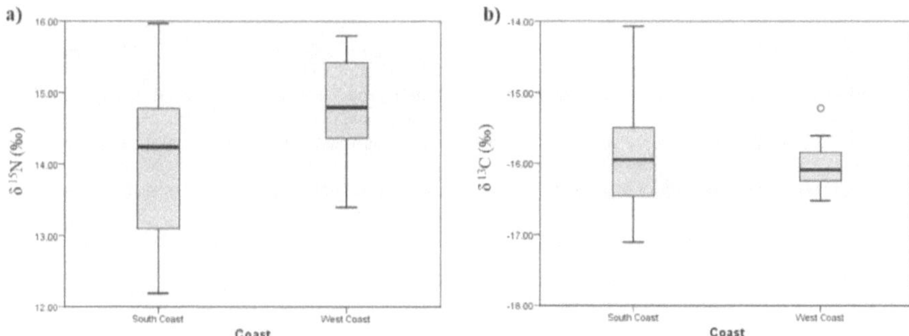

Figure 40: The range of isotopic signatures, expressed by delta notation in parts per thousand (‰), of ^{15}N (δ ^{15}N; a) and ^{13}C (δ^{13}C; b), for *Squalus bassi* sharks sampled on the South and West coast of South Africa. Significant differences were deemed at $p <0.05$, and were evident between the coasts for δ^{15}N ($H' = 16.26$; $df = 1$; $p <0.001$) and δ^{13}C ($F_1 = 5.57$, $df = 1$; $p <0.05$). The boxplot provides a five number summary of the dataset, and depicts the minimum and maximum values of the dataset (lower and uppermost lines), the 25th and 75th percentile (bottom and top most inner lines) and the median (horizontal line through the center of the box). Outliers are represented by circles (°).

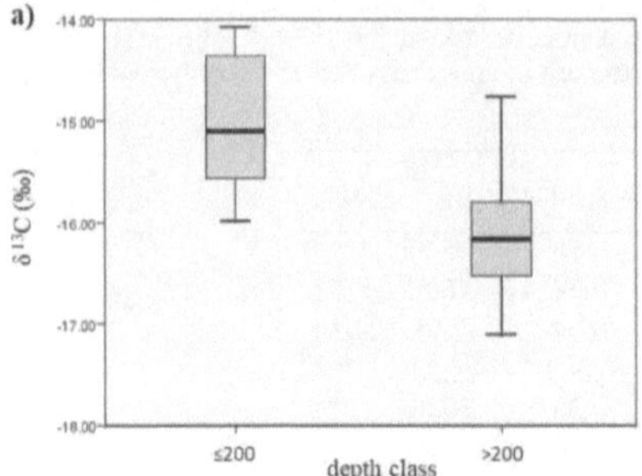
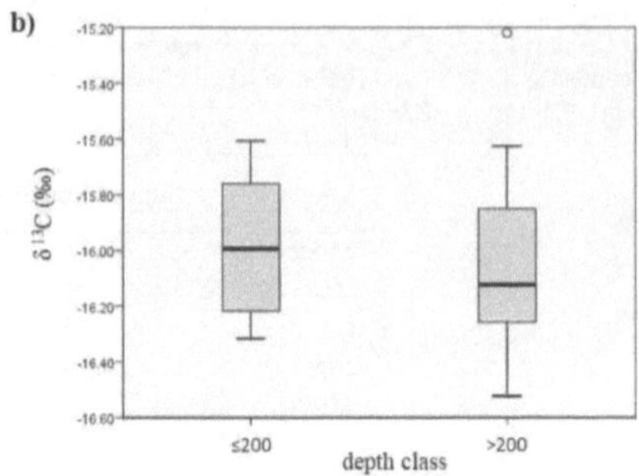

Figure 41: The range of isotopic signatures, expressed by delta notation in parts per thousand (‰), of ^{13}C (δ ^{13}C), for *Squalus bassi* sharks sampled at depths of ≤ 200 m and > 200 m on the South (**a**) and West coast (**b**) of South Africa. Significant differences were deemed at $p < 0.05$, and were evident between the two depths on the same coast ($F_1 = 19.86$, $df = 1$; $p < 0.001$). The boxplot provides a five number summary of the dataset, and depicts the minimum and maximum values of the dataset (lower and uppermost lines), the 25th and 75th percentile (bottom and top most inner lines) and the median (horizontal line through the center of the box). Outliers are represented by circles (°).

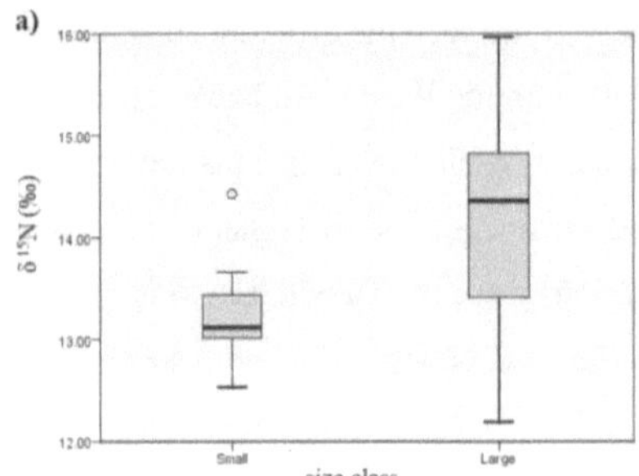
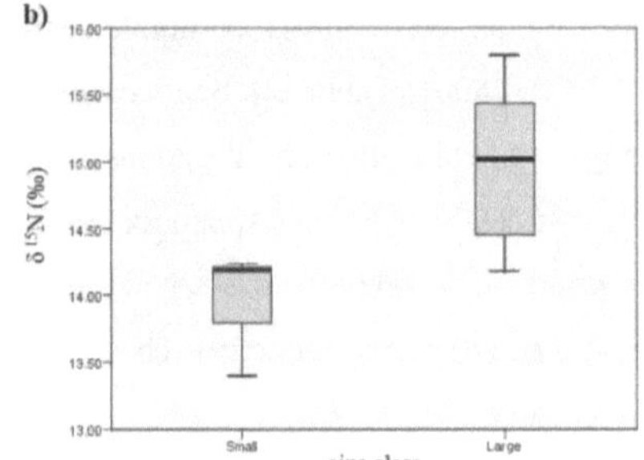

Figure 42: The range of isotopic signatures, expressed by delta notation in parts per thousand (‰), of ^{15}N (δ ^{15}N), for small and large *Squalus bassi* sharks sampled on the South (**a**) and West coast (**b**) of South Africa. Significant differences were deemed at $p < 0.05$, and were evident between the sizes on each coast (West Coast: $H' = 6.96$; $df = 1$; $p < 0.05$; South Coast: $H' = 8.01$; $df = 1$; $p < 0.05$). The boxplot provides a five number summary of the dataset, and depicts the minimum and maximum values of the dataset (lower and uppermost lines), the 25th and 75th percentile (bottom and top most inner lines) and the median (horizontal line through the center of the box). Outliers are represented by circles(°).

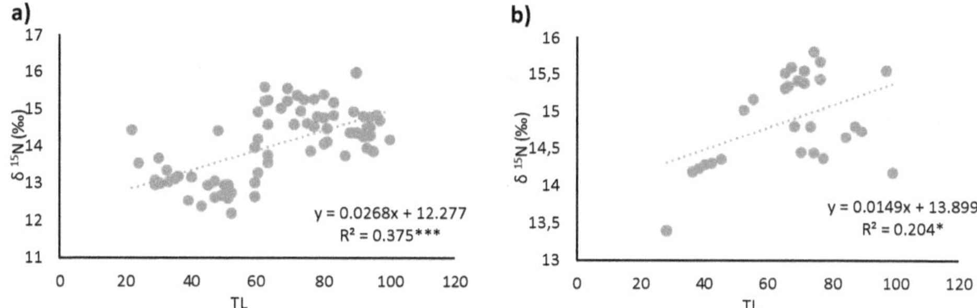

Figure 43: The scatterplot depicting the isotopic signature, expressed by delta notation in parts per thousand (‰), of ^{15}N (δ ^{15}N) vs size (TL; cm), for *Squalus bassi* sharks sampled on the South (**a**) and West coast (**b**). Both graphs include an associated linear trendline equation and the R^2 value. Significance is deemed at $p < 0.05$ and the level of significance of the correlation is indicated by the use of an asterisk (i.e. $p < 0.05$ *; $p < 0.01$ **; $p < 0.001$ ***; $p < 0.0001$ ****).

Interspecific variability in δ¹⁵N and δ¹³C of Squalus acutipinnis and S. bassi

From the isotopic biplot, a leftward clustering of West coast samples is seen with South coast samples displaying a more widespread range with overlap, rather than separation being apparent (Fig. 44).

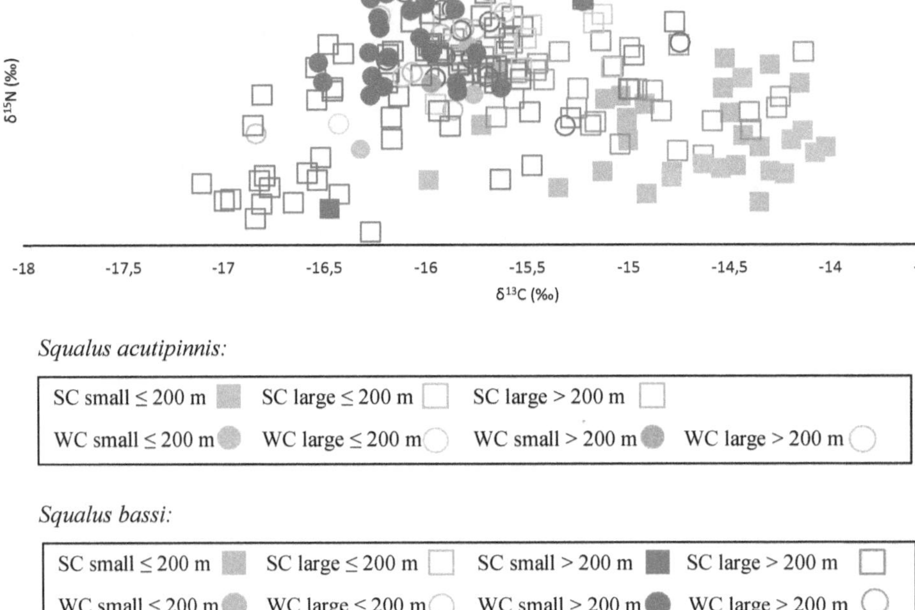

Squalus acutipinnis:

SC small ≤ 200 m	SC large ≤ 200 m	SC large > 200 m	
WC small ≤ 200 m	WC large ≤ 200 m	WC small > 200 m	WC large > 200 m

Squalus bassi:

SC small ≤ 200 m	SC large ≤ 200 m	SC small > 200 m	SC large > 200 m
WC small ≤ 200 m	WC large ≤ 200 m	WC small > 200 m	WC large > 200 m

Figure 44: The isotopic signatures, expressed by delta notation in parts per thousand (‰), of ¹⁵N (δ ¹⁵N) and ¹³C (δ¹³C), for small (filled) and large (empty) *Squalus acutipinnis* sharks, sampled at depths of ≤ 200 m (green) and > 200 m (blue) on the South (SC: square) and West (WC: circle) Coast as well as for small (filled) and large (empty) *Squalus acutipinnis* sharks, sampled at depths of ≤ 200 m (orange) and > 200 m (red) on the South (SC: square) and West (WC: circle) coast of South Africa.

Overall, *Squalus bassi* had lower average δ¹³C values than *S. acutipinnis* on both coasts (Fig 45a and 45b; Appendix 29). Overall, at depths ≤ 200 m on both the South and West coast, there were no differences between the two species in either δ¹⁵N or δ¹³C. However, there were significant differences between species at depths > 200 m. On the South coast, *S. acutipinnis* had higher δ¹⁵N and δ¹³C values than *S. bassi* (Fig. 46a and 46b; Appendix 29), whereas *S. bassi* had significantly lower δ¹³C values than *S. acutipinnis* on the

West coast (Fig. 47; Appendix 29). Interestingly, there were significant differences in δ^{15}N between small sharks of each species on the South, but not West coasts. Small *S. acutipinnis* had a higher average δ^{15}N value than *S. bassi* (Fig. 48; Appendix 29). For large sharks, on both coasts, only differences in δ^{13}C values were evident (Fig. 49a and 49b). *Squalus bassi* had lower δ^{13}C values than *S. acutipinnis* on both coasts (Fig. 49a and 49b; Appendix 29).

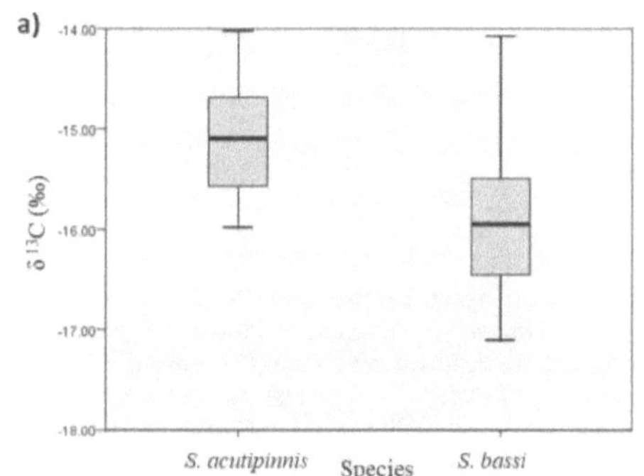
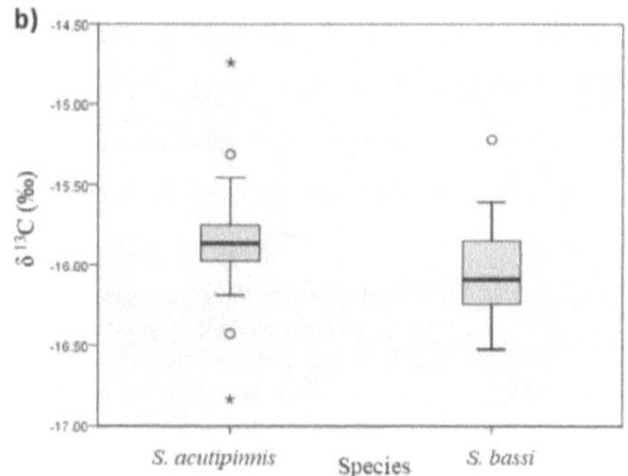

Figure 45: The isotopic range, expressed by delta notation in parts per thousand (‰), of ^{13}C (δ^{13}C), for *Squalus acutipinnis* and *S. bassi* sampled on the (**a**) South and (**b**) West coast of South Africa. Significant differences were deemed at p <0.05, and were evident between species on each coast (SC: H' = 37.85; df = 1; p <0.001; WC: H' = 6.36; df = 1; p <0.05). The boxplot provides a five number summary of the dataset, and depicts the minimum and maximum values of the dataset (lower and uppermost lines), the 25th and 75th percentile (bottom and top most inner lines) and the median (horizontal line through the center of the box). Outliers are represented by an asterisk (*).

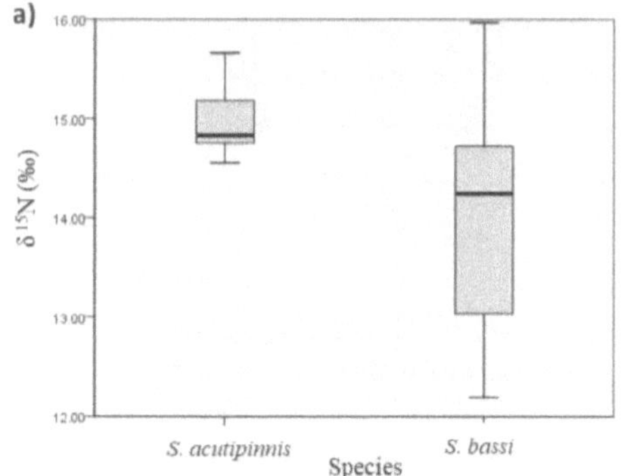
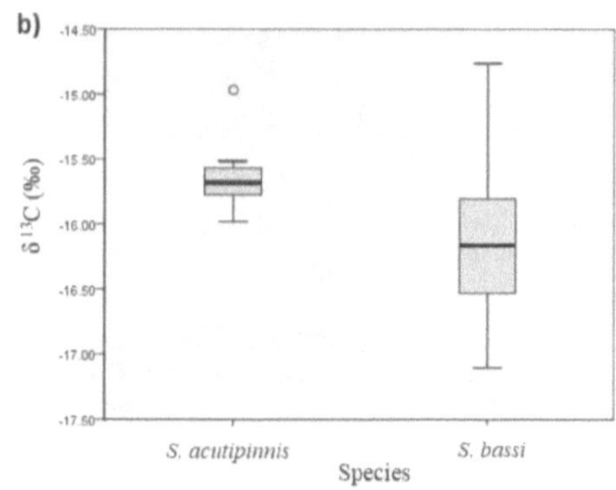

Figure 46: The range of isotopic signatures, expressed by delta notation in parts per thousand (‰), of ^{15}N (δ^{15}N; **a**) and ^{13}C (δ^{13}C; **b**), for *Squalus acutipinnis* and *S. bassi* sampled at depths > 200 m on the South coast of South Africa. Significant differences were deemed at p <0.05, and were evident between species at this depth on this coast (δ^{15}N: H' = 14.30; df = 1; p <0.001; δ^{13}C: H' = 11.93; df = 1; p <0.001). The boxplot provides a five number summary of the dataset, and depicts the minimum and maximum values of the dataset (lower and uppermost lines), the 25th and 75th percentile (bottom and top most inner lines) and the median (horizontal line through the center of the box). Outliers are represented by an asterisk (*).

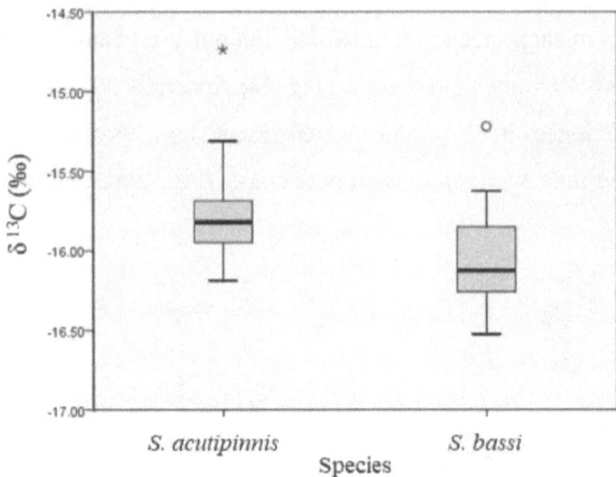

Figure 47: The range of isotopic signatures, expressed by delta notation in parts per thousand (‰), of ^{13}C (δ ^{13}C), for *Squalus acutipinnis* and *S. bassi* sampled at depths > 200 m on the West coast of South Africa. Significant differences were deemed at p <0.05, and were evident at this depth on this coast (H' = 9.62; df = 1; p <0.05). The boxplot provides a five number summary of the dataset, and depicts the minimum and maximum values of the dataset (lower and uppermost lines), the 25th and 75th percentile (bottom and top most inner lines) and the median (horizontal line through the center of the box). Outliers are represented by an asterisk (*).

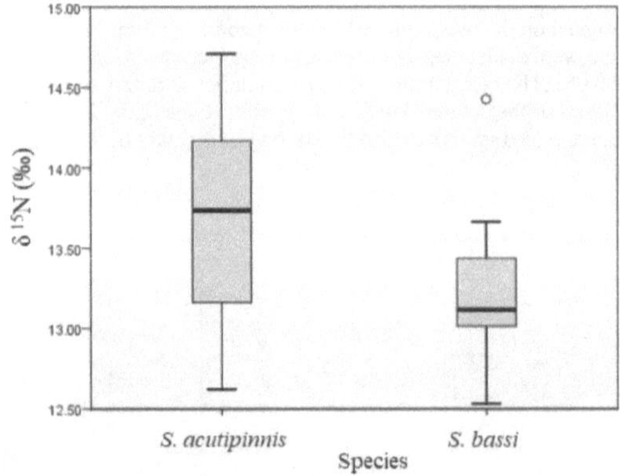

Figure 48: The range of isotopic signatures, expressed by delta notation in parts per thousand (‰), of ^{15}N (δ ^{15}N), for small *Squalus acutipinnis* and *S. bassi* sampled on the South coast of South Africa. Significant differences were deemed at p <0.05, and were evident between species of this size on this coast (H' = 3.94; df = 1; p <0.05). The boxplot provides a five number summary of the dataset, and depicts the minimum and maximum values of the dataset (lower and uppermost lines), the 25th and 75th percentile (bottom and top most inner lines) and the median (horizontal line through the center of the box). Outliers are represented by an asterisk (*).

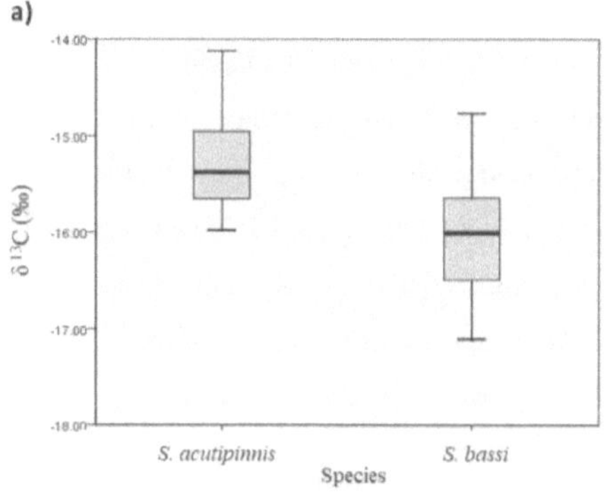
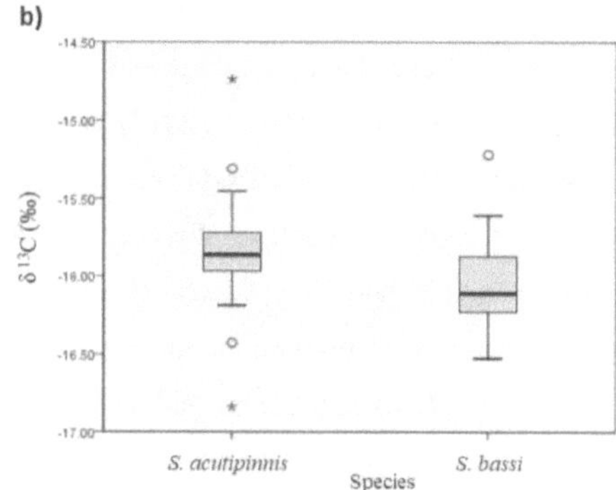

Figure 49: The range of isotopic signatures, expressed by delta notation in parts per thousand (‰), of ^{13}C (δ ^{13}C), for large *Squalus acutipinnis* and *S. bassi* sampled on the South (**a**) and West (**b**) coast of South Africa. Significant differences were deemed at p <0.05, and were evident between species of this size on each coast (SC: *H'* = 39.80; *df* = 1; p <0.001; WC: *H'* = 6.44; *df* = 1; p <0.05). The boxplot provides a five number summary of the dataset, and depicts the minimum and maximum values of the dataset (lower and uppermost lines), the 25th and 75th percentile (bottom and top most inner lines) and the median (horizontal line through the center of the box). Outliers are represented by an asterisk (*).

Discussion

For both species, δ^{15}N was significantly greater and δ^{13}C was significantly lower on the West coast than the South coast. Similar West vs South coast differences have been previously reported around South Africa (Hill *et al.*, 2006; Hill and McQuaid, 2008; van der Lingen and Miller, 2014; van der Heever, 2017). For example, Parkins, (1993) noted that *Merluccius capensis* on the West coast had greater δ^{15}N signatures than those on the South coast; an observation supported some 20-years later by van der Lingen and Miller (2014) on the same species. Van der Heever (2017) has recently shown this pattern for two catsharks in the same region.

Sharks on the West coast may be eating higher-trophic-level prey than on the South coast and such would explain the more enriched δ^{15}N values noted. This line of reasoning is supported by the results of the stomach content analyses (Chapter Three), which indicate that arthropods/crustaceans were more commonly eaten on the South coast, whilst teleosts were more important on the West coast. Whether this reflects localised changes in the availability of prey (Meyer and Smale, 1991; Rinewalt, 2007; Woodland *et al.*, 2011; van der Heever, 2017) or regional differences in feeding behaviours (Andrews, 2010) is unknown.

Alternatively, spatial differences in the $\delta^{15}N$ of dogfish around South Africa may simply reflect differences in regional isotopic baselines (Post, 2002; van der Lingen and Miller, 2014; van der Heever, 2017). Regional differences in $\delta^{15}N$ have previously been associated with and described to be affected by both the source and type of nitrogen that is used in primary production (Waser *et al.*, 2000; Montoya *et al.*, 2002; Post, 2002; van der Lingen and Miller, 2014; van der Heever, 2017). Upwelling on the West Coast injects new nitrogen in the form of nitrate into the system and this carries a higher $\delta^{15}N$ signal than recycled or N-fixed nitrogen, and the latter may be more prevalent on the South coast (Probyn *et al.*, 1994; Montoya *et al.*, 2002; Waite *et al.*, 2007; van der Lingen and Miller, 2014).

Two other explanations can be invoked to explain the spatial differences in $\delta^{15}N$, though neither have convincing support. Firstly, it has also been shown that long periods between meals results in higher $\delta^{15}N$ signatures (Hobson *et al.*, 1993; Gannes *et al.*, 1997; Reid *et al.*, 2013). Whilst it is possible that individual fish on the West coast may consume prey items less frequently than those on the South coast, either because prey is less abundant and/or is of a higher nutritional value, collectively 28.5% of the stomachs on the West coast was empty while only 25% of the stomachs sampled on the South coast was empty (Chapter Three). Alternatively, because the two coasts were sampled during different seasons (South coast during spring, West coast during summer), changes in $\delta^{15}N$ could reflect differences in seasonal tissue-turn over rates (Blanco-Parra *et al.*, 2012; Wyatt *et al.*, 2012; Heithaus *et al.*, 2013; Reum and Essington, 2013). However, previous research on elasmobranch trophic ecology has shown that this is an unlikely cause, as the rate that tissues turnover in elasmobranchs species is slow (Domi *et al.*, 2005; Heithaus *et al.*, 2013).

The $\delta^{13}C$ value is regarded as reflecting the source of primary productivity (Post, 2002; Domi *et al.*, 2005). Macrophytes, macroalgae and phytoplankton have different $\delta^{13}C$ signatures, which reflect a decreasing gradient of distance from shore, with macrophytes and macroalgae being more enriched than phytoplankton (Hill *et al.*, 2006; Hill and McQuaid, 2008). The benthic environment is said to be $\delta^{13}C$-enriched relative to the pelagic environment, and a nearshore – offshore depletion of $\delta^{13}C$ has frequently been shown (France, 1995; Hobson *et al.*, 1996; France and Peters, 1997; Hobson, 1999; Domi *et al.*, 2005; Miller *et al.*, 2008; Shipley *et al.*, 2017).

The tissues of both species were more depleted in $\delta^{13}C$ on the West than on the South coast. This could indicate that sharks on the West coast had a preference for prey of low $\delta^{13}C$, or that prey species of lower $\delta^{13}C$ signatures were more readily available (van der Lingen and Miller, 2014). The enriched $\delta^{13}C$ noted on the South coast might be as a consequence of pronounced benthic-pelagic coupling that occurs there (van der Lingen and Miller, 2014). It should be remembered that sharks on the South coast consumed substantial proportions of benthic invertebrates that may rely on the microphyto-benthic plankton as their major source of production, and these are known to be enriched in $\delta^{13}C$ (McConnaughey and McRoy, 1979; France, 1995). Alternatively, as most of the samples collected on the West coast were collected at deeper/offshore locations than on the South coast, differences may simply reflect the natural onshore-offshore depletion of $\delta^{13}C$ (Post, 2002; Hill *et al.*, 2006; Post *et al.*, 2007; Hill and McQuaid, 2008; van der Lingen and Miller, 2014).

Moreover, the result of lower $\delta^{13}C$ signals on the West coast, rather than South coast is particularly interesting, as an ecosystem dominated by a diatom phytoplankton community should be $\delta^{13}C$-enriched (Vander Zanden and Vadeboncoeur, 2002; van der Lingen and Miller, 2014). Hill *et al.* (2006) noted a similar spatial change in $\delta^{13}C$ around South Africa, which they attributed to differences in oceanography. As previously noted, the Benguela Current brings cool, nutrient-rich waters onto the shelf, and cold water is known to increase CO_2 solubility and therefore to promote $\delta^{13}C$-depletion (Arthur *et al.*, 1985; Rau *et al.*, 1989; Fry, 1996; Hill *et al.*, 2006).

Changes in $\delta^{15}N$ and/or $\delta^{13}C$ signatures with changes in depth have previously been noted for many marine species (Hill *et al.*, 2006; Hill and McQuaid, 2008; Andrews, 2010; Hussey *et al.*, 2012; Churchill *et al.*, 2015; Shipley *et al.*, 2017; van der Heever, 2017). Significantly greater values of $\delta^{15}N$ were noted for *Squalus acutipinnis* at the deep (> 200 m), rather than shallow depths ($\leq$ 200 m) on both coasts. On the other hand, sharks of both species, displayed significantly lower $\delta^{13}C$ signatures at the deeper depths on the South coast, with the same pattern only serving true for *S. bassi* on the West coast.

Such changes may reflect a diet change with depth, particularly relating to prey availability and/or feeding location (Michener and Schell, 1994; Michener and Kaufman, 2007; Hussey *et al.*, 2012; Ebert *et al.*, 2015; van der Heever, 2017). Changes in $\delta^{15}N$ and $\delta^{13}C$ values, with depth, could also be attributed to inshore - offshore changes in both trophic pathways and/or baseline isotopic values (Parkins, 1993; Post, 2002; Andrews, 2010; Hussey *et al.*, 2012; Shipley *et al.*, 2017). Prey consumed by these sharks, in the shallows, may have lower δ^{15}N and higher δ^{13}C values than those prey consumed at great depths. Literature, however, indicates that prey inhabiting inshore/shallow ecosystems rely on microphyto-benthic sources of production, which are enriched in both isotopes of δ^{15}N and δ^{13}C, and should therefore consequently house prey that is enriched in these isotopes (McConnaughey and McRoy, 1979; France, 1995). If so, this explanation only supports the enriched δ^{13}C signatures observed in the shallow by this study as δ^{15}N signatures became enriched with depth.

The δ^{15}N values of these sharks might be as a consequence of the "bigger-deeper" phenomenon that exists in many species of elasmobranchs (chapters two and three) (Domi *et al.*, 2005; Hussey *et al.*, 2012; Kiszka *et al.*, 2014; Churchill *et al.*, 2015; Shipley *et al.*, 2017; van der Heever, 2017). The bigger sharks found in the deep tend to consume bigger prey, of higher trophic status and thus may result in them having significantly higher δ^{15}N signatures in the deep (Domi *et al.*, 2005; Andrews, 2010; van der Heever, 2017). The results of the stomach content analyses provide support for this as with increased depth, an increase in predator size, as well as prey size consumed, was noted (Chapter Three).

The source of primary production available to the sharks of the deep may also provide an understanding as to why these sharks have significantly lower δ^{13}Csignatures. Shipley *et al.* (2017) explained that the overlaying marine phytoplankton of the pelagic environment, which is naturally δ^{13}C-depleted, become available to deep-sea sharks in the form of marine snow, through extensive diel vertical migrations performed by the pelagic/mesopelagic community. These migrations transport nutrients from surface waters to deep waters and may then be, consequently, influencing the values of the isotopes of the deep (Polunin *et al.*, 2001; Trueman *et al.*, 2014; Churchill *et al.*, 2015; Shipley *et al.*, 2017). Studies by Takai *et al.* (2002) and Kadila, (2019) have both described decreases in δ^{13}C signatures with depth, for the species they have studied.

As depth increase with distance from shore, δ^{13}C signatures have been shown to deplete, a pattern that has also previously been described for various marine organisms around South African coastline (Hill *et al.*, 2006; Hill and McQuaid, 2008). Herein, the decreased influence of nearshore carbon (macrophytes) with distance from shore has been attributed as a likely reason (Hill *et al.*, 2006). Moreover, deeper water is colder and cold water results in increased solubility of carbon dioxide which causes $\delta^{13}C$ depletion (Arthur *et al.*, 1985, Fry, 1996, Hill *et al.*, 2006).

Body size is an important factor to consider in trophic ecology studies, due to its influence over the outcomes of predator-prey interactions, predator foraging capabilities and metabolic processes (Cohen *et al.*, 1993; Brown, 2004; Collins *et al.*, 2005; Andrews, 2010; Reid *et al.*, 2013). For both species and on both coasts, $\delta^{15}N$ signatures were significantly greater in the large sharks, while a significant (negative) effect in $\delta^{13}C$ signatures was only true for *Squalus acutipinnis* on the South coast.

The trend of larger organisms being more enriched in their $\delta^{15}N$ signatures than their small counterparts is commonly reported for marine species and appears related to ontogenetic shifts in diet (Davenport and Bax, 2002; Estrada *et al.*, 2006; Newsome *et al.*, 2010; Kinney *et al.*, 2011; Vaudo and Heithaus, 2011; Courtney and Foy, 2012; Kim *et al.*, 2012; Reid *et al.*, 2013; Shipley *et al.*, 2017; van der Heever, 2017). A bigger body allows for increased swim speeds, greater endurance while searching for prey, better prey capturing and handling abilities, with a bigger gape size allowing for the consumption of large prey, that will be of greater energy value (Pyke, 1984; Tilley *et al.*, 2013; Kiszka *et al.*, 2014; Tillett *et al.*, 2014; van der Heever, 2017). Larger sharks may feed on larger prey, which are consequently of high trophic status, thus resulting in enriched $\delta^{15}N$ signals (Hussey *et al.*, 2012; Churchill *et al.*, 2015; Shipley *et al.*, 2017). The results of the stomach content analyses support this, as larger sharks fed on larger sized prey, while also showing an increase in the diversity of prey consumed (Chapter Three). As a further example, Tilley *et al.* (2013) studied the trophic and movement ecology of mesopredators in the Caribbean, by using SIA in conjunction with acoustic tracking telemetry. These authors found that stingrays did not change their habitat as they grew, but rather fed on larger prey as well as increased the diversity of prey they consumed, which resulted in their enriched $\delta^{15}N$ signatures.

In addition to this, and as earlier alluded to, the body size of organisms is influenced by bathymetry, and bigger animals have been shown to feed and occur in the deep (Domi *et al.*, 2005; Shipley *et al.*, 2017; van der Heever, 2017). Moreover, the quality as well as the quantity of food in the deep-sea is said to be influenced by the overlaying productivity, which too was earlier alluded (Estrada *et al.*, 2006; Kadila, 2019). The depleted $\delta^{13}C$ signature with size, as seen in this study may also relate to the bathymetric effects. When sharks are small, they inhabit the inshore and shallow benthic habitats that are enriched in $\delta^{13}C$ and as theygrow, they may be performing offshore migration, and larger sharks remain in deeper waters, where δ^{13}C-depleted prey become incorporated into their diet (Sherwood and Rose, 2005; Shipley *et al.*, 2017). Hussey *et al.* (2011) analysed the SI profiles for multiple marine top predator species and sought to, among other things, validate this method in terms of its usefulness as a predictor for trophic positioning. One general trend that these authors found was that as certain species grew, they showed a depleted δ^{13}C signal. Here too, these authors related this with the fact that the larger animals spent lots of time foraging in the deep offshore waters off the shelf edge, where waters are typically δ^{13}C-depleted.

Interspecific variability in the stable isotope signatures of δ^{15}N and δ^{13}C was shown by region, depth and size; however differences were not common. A significant difference in the δ^{13}C signals was evident between species on each coast, with *Squalus bassi* having much lower δ^{13}C signatures than *S. acutipinnis*. This may be as a result of *S. bassi* being larger as well as mostly being sampled in much deeper waters than *S. acutipinnis*.

Significant differences at depths ≤ 200 m were evident on the South coast while significant differences in just δ^{13}C signatures were evident between species on the West coast, at depths > 200 m. At depths ≤ 200 m, *Squalus acutipinnis* was shown to be more enriched in δ^{15}N, whilst *S. bassi* were more depleted in δ^{13}C. As *S. bassi* are likely found at the depth extremes of this depth bin and are the biologically bigger shark, these results would be expected. However, the fact that *S. acutipinnis* is more enriched in δ^{15}N signature may relate to the fact that more samples of *S. acutipinnis* were collected at this depth bin on the South coast, and may just be an indication of possible bias.

On the South coast, variability with size was shown between small *Squalus acutipinnis* and *S. bassi* based on their δ^{15}N signature, as well as between large sharks based on their δ^{13}C signatures. Small *S. acutipinnis* were shown to have significantly higher δ^{15}N signatures when compared to small *S. bassi*. There were more small *S. acutipinnis* sampled than small *S. bassi*, thus *S. bassi* had a much smaller data pool. This result may thus reflect the unbalanced sample collection. The fact that large *S. bassi* were significantly more depleted in δ^{13}C too, might be as a result of these sharks occurring deeper and are biologically bigger than large *S. acutipinnis*. When comparing diets by stomach content analysis for this study, it is clear that large *S. bassi* focused largely on consuming big prey and fed mostly on teleosts and cephalopods whereas although *S. acutipinnis* fed on a substantial amount of teleosts and cephalopods, their diets still comprised of lots of benthic invertebrates. Peterson, (2014) showed similar results when assessing the trophic ecology of various teleosts and elasmobranchs in Florida Big Bend.

This study provided the first attempt at understanding intraspecific and interspecific variability in the isotope signatures of $\delta^{15}N$ and $\delta^{13}C$ by *Squalus acutipinnis* and *S. bassi* in the West and South coast waters of South Africa. The results indicated that there was substantial intraspecific and interspecific overlap in isotopic signatures and significant variability was not widespread. Species may then either, be feeding on the same prey, or even in the same space. However, when assessing SCA, it is clear that although the same prey types were being consumed by these sharks, they were consumed in varying proportions, which may indicate possible resource partitioning.

Moreover, this study has proved particularly useful in its assessment of the observed variation in isotopic signatures, which may be related to various oceanographic, biological and ecological factors that these sharks may be faced with, some of which include biogeographic shifts in isotopic baselines, onshore-offshore gradients, feeding location and habit as well as trophodynamics. However, as much as it proved useful, the study of isotope ecology is not without limitations.

Sample collection was unbalanced and only covered trawlable regions. Moreover, although stable isotope results indicate assimilated prey, it does not provide information on exact prey type consumed. This makes direct diet comparisons difficult. Therefore, to improve the results, more rigorous sampling should occur, that covers greater habitat expanses and is not just limited to trawlable areas Importantly, an attempt at acquiring, equal across-board samples should be prioritised, so that comprehensive coast, depth and size class comparisons could be made, both intraspecifically and interspecifically. In addition to this, the SIA study should include the signatures of various prey species, to get a more holistic view of what is likely happening in the ecosystem. Basal prey signatures should also be determined as this will, in turn, provide better insight into the differing isotopic baselines presented on these coasts, from which biogeographic shifts may be inferred. Multiple tissue comparisons should also occur, which will work to assess varying temporal variations. Furthermore, a future stable isotope ecological study should be used in conjunction with both SCA and acoustic telemetry as it will then provide a time-integrated dataset. Even though this study was the first attempt at understanding intraspecific and interspecific variability in the isotope signatures of $\delta^{15}N$ and $\delta^{13}C$ by *S. acutipinnis* and *S. bassi*, and is not without limitations, it provides a useful foundation for future work.

Chapter Five

General Conclusions

This study investigated intraspecific and interspecific variability in the distribution and feeding ecology of two common and co-occurring dogfish species, *Squalus acutipinnis* and *S. bassi* that are both common bycatch species of the demersal fishery around South Africa. Data indicate that the two species are not randomly distributed (Chapter Two) and that *S. acutipinnis* were more commonly caught in the trawls on the South coast than *S. bassi*, whereas *S. bassi* were more commonly caught in the trawls on the West coast than *S. acutipinnis*, with significant intraspecific differences in the overall distribution by depth and size being further evident. Notably, shark size increased with depth. In addition to this, sexual differences were highlighted, with females of both species being more abundant on the South coast and males on the West coast.

Given general similarities in morphology and presumed diet, these differences in distribution imply a degree of habitat partitioning (by coast, sex, depth of capture and size class), both intraspecific and interspecific, which may facilitate and promote co-existence. Results of this nature are not unlikely as similar findings have been reported in numerous other studies investigating the distribution of sympatric elasmobranchs around the world (Platell *et al.*, 1998; White and Potter, 2004; Papastamatiou *et al.*, 2006, 2009; Pita *et al.*, 2011; O'Shea, 2012; van der Heever, 2017). It should however be remembered that in order for competition to exist, resources must be limited (Schoener, 1974; Pianka, 1981; Carrassón *et al.*, 1992). Therefore, alternative reasons for the observed distributions should be considered. For example, size segregation by depth may be as a result of ontogenetic shifts or even as an act to counter predation of conspecifics.

Although competition is a strong means by which species select certain habitats (Broennimann *et al.*, 2012), physiological responses and resource abundances also play a big role in habitat selection (Knip *et al.*, 2010; Taylor *et al.*, 2011). The distribution of these sharks may therefore also be influenced by the differing oceanographic and environmental conditions present on the West and South coast (Lutjeharms, 2006; Hutchings *et al.*, 2009). If this is true, the observed distributions may not be as a result of competition, but rather as a result of the environmental tolerance of each species (van der Heever, 2017).

Analysis of stomach contents shows both separation and overlap in dietary resources among and between the two species (Chapter Three). Teleosts, crustaceans, cephalopods, and polychaetes were the prey categories of significant importance in the diets of these sharks with the degree of importance depending on species, coast, depth and size class. Sex had a negligible effect on diet. Clear differences in the diet between coasts were evident and likely reflect changes in prey availability between the coasts. Similar results have been shown for many other elasmobranch species (Cortés and Gruber, 1990; Yamaguich and Taniuchi, 2000; Heithaus, 2001; Bethea *et al.*, 2006).

Even though these sharks are shown to have generalist (perhaps opportunistic) dietary habits, significant intraspecific variability in their diet that may be related to ontogenetic shifts in the diet with size/depth was observed. Bigger sharks move into deep water where they likely encounter larger and more nutritious meals: teleosts become more important in the diets of larger fish of both species. Intraspecific changes in diet by depth and size are common in squaloids and are related to ontogeny (Kazunari, 1991; McMillan *et al.*, 1999; Braccini *et al.*, 2005; Bigman, 2013). Interspecific variability in the diets of these sharks was apparent, with co-occurring sharks feeding on the same dietary resources, but in differing proportions. Although these differences may reflect partitioning, other differences may also mirror spatial differences in prey-type availability, when species do not co-occur. Habitat separation by species/depth/size class may also be responsible for driving variability and this has been previously reported for many elasmobranch species (Carrassón *et al.*, 1992; Flammang *et al.*, 2011; van der Heever, 2017).

There were significant intraspecific and interspecific differences in the stable isotope composition of the two species linked to changes in coast, depth occupied and size class evaluated (Chapter Four). The δ^{15}N signatures of both species were significantly greater, while their δ^{13}C signature was significantly lower on the West coast than the South coast. This may be attributed to regional differences in isotopic baselines, as noted for other species (Parkins, 1993; Hill *et al.*, 2006; van der Lingen and Miller, 2014; van der Heever, 2017). The enriched δ^{13}C on the South coast might be as a consequence of pronounced benthic-pelagic coupling that occurs there (van der Lingen and Miller, 2014) or because as most tissue samples collected on the West coast were collected at deeper/offshore depth, thus differences may even reflect the onshore-offshore depletion of δ^{13}C (Post *et al.*, 2002; Hill *et al.*, 2006). Differences by depth were also evident as sharks occurring deeper displayed higher δ^{15}N and lower δ^{13}C signatures, with the opposite being true for sharks of the shallows. In general, smaller sharks showed lower δ^{15}N and higher δ^{13}C signatures than large sharks. Here, the ontogenetic shifts in diet, with size and depth occupation does come into play as bigger sharks occur deeper than smaller sharks. Moreover, bigger sharks may be eating bigger prey of higher $\delta^{15}N$ status, while the lower $\delta^{13}C$ signatures may simply be attributed to the natural onshore-offshore gradient of $\delta^{13}C$, with distance from shore.

This study has provided various lines of evidence in the support for intra and interspecific variability in the distribution and feeding ecology of *Squalus acutipinnis* and *S. bassi* around the West and South coasts of South Africa. Interpreting the results in the context of competition alone is difficult, as there is little evidence that resources are limited within the environment (Schoener, 1974; Pianka, 1981; Ross, 1986) and, as with many other studies, the limitations of the data collected needs to be recognised.

Further studies are recommended because these endemics are a major bycatch in this region and as commercial fisheries go deeper they will likely exploit greater proportions of their populations (Clarke *et al.*, 2005; Ferretti *et al.*, 2010; Pajuelo *et al.*, 2011; Oliver *et al.*, 2015). Cascading effects of the removal of apex and mesopredators have been widely reported (Stevens *et al.*, 2000; Baum and Worm, 2009; Ferretti *et al.*, 2010, 2013; Dulvy *et al.*, 2014; Heupel *et al.*, 2014) and having comprehensive ecological knowledge of these species will allow for risk assessments to be undertaken that can contribute to fisheries management (Simpfendorfer and Kyne, 2009; Mourier *et al.*, 2012; Schoener *et al.*, 2012; Peterson, 2014; Navarro *et al.*, 2016). We should be using a larger source of both fisheries dependent and independent data, while also incorporating data from various fishing methods, which will allow us access to more information from untrawlable grounds – thus ensuring that the lack of sampling over untrawlable grounds are not a source of bias in the studies of this nature.

We need to improve the resolution of diet and should aim to equalise sample sizes across all categories (species by coast, depth, size, sex), with at least 200 stomachs per category (Morata *et al.*, 2003, Bracinni *et al.*, 2005). Complementary work on stable isotopes should be undertaken that is similarly balanced and category inclusive. Information on prey signatures will provide insights into the differing isotopic baselines that are presented on both coasts, from which biogeographic shifts can then be inferred.

References

Acuña, E. & Villarroel, J. C. 2010. Feeding habits of two deep-sea sharks from central-northern Chile: hooktooth dogfish *Aculeola nigra* (Etmopteridae) and dusky catshark *Bythalaelurus canescens* (Scyliorhinidae). Rev. Biol. Mar. Oceanogr. 45(1), 737–743.

Alonso, M. K., Crespo, E. A., García, N. A., Pedraza, S. N., Mariotti, P. A., Berón Vera, B. & Mora, N. J. 2001. Food habits of *Dipturus chilensis* (Pisces: Rajidae) off Patagonia, Argentina. ICES J. Mar. Sci. 58(1), 288–297.

Alonso, M. K., Crespo, E. A., Mariotti, P. A. & Mora, N. J. 2002. Fishery and ontogenetic driven changes in the diet of the spiny dogfish, *Squalus acanthias*, in Patagonian waters, Argentina. Environ. Biol. Fishes 63(2), 193–202.

Andrews, A. 2010. Variation in the trophic position of spiny dogfish (*Squalus acanthias*) in the northeastern Pacific Ocean: An approach using carbon and nitrogen stable isotopes. MSc book, University of Alaska Fairbanks, United States.

Arthur, M., Dean, W. & Claypool, G. 1985. Anomalous ^{13}C enrichment in modern marine organic carbon. Nature 315(6016), 216–218.

Aschliman, N., Nishida, M., Miya, M., Inoue, J., Rosana, K. & Naylor, G. 2012. Body plan convergence in the evolution of skates and rays (Chondrichthyes: Batoidea). Mol. Phylogenet. Evol. 63(1), 28–42.

Attwood, C., Petersen, S. & Kerwath, S. 2011. Bycatch in South Africa's inshore trawl fishery as determined from observer records. ICES J. Mar. Sci. 68(10), 2163–2174.

Badenhorst, A. & Smale, M. J. 1991. The distribution and abundance of seven commercial trawlfish from the Cape south coast of South Africa, 1986–1990. S. Afr. J. Mar. Sci. 11(1), 377–393.

Baker, R., Buckland, A. & Sheaves, M. 2014. Fish gut content analysis: Robust measures of diet composition. Fish Fish. 15(1), 170–177.

Bangley, C. & Rulifson, R. 2014. Observations on spiny dogfish (*Squalus acanthias*) captured in late spring in a North Carolina estuary. F1000Research 3(189), 1–8.

Barausse, A., Correale, V., Curkovic, A., Finotto, L., Riginella, E., Visentin, E. & Mazzoldi, C. 2014. The role of fisheries and the environment in driving the decline of elasmobranchs in the northern Adriatic Sea. ICES J. Mar. Sci. 71(7), 1593–1603.

Barbini, S. A., Lucifora, L. O. & Hozbor, N. M. 2011. Feeding habits and habitat selectivity of the shortnose guitarfish, *Zapteryx brevirostris* (Chondrichthyes, Rhinobatidae), off north Argentina and Uruguay. Mar. Biol. Res. 7(4), 365–377.

Bass, A., Aubrey, J. & Kistnasamy, N. 1976. Sharks of the east coast of southern Africa. VI. The families Oxynotidae, Squaloidae, Dalatiidae, and Echinorhinidae. Invest. Rep. Oceanogr. Res. Inst. 45(1), 103.

Baum, J. K. 2003. Collapse and Conservation of Shark Populations in the Northwest Atlantic. Science 299(5605), 389–392.

Baum, J. K. & Worm, B. 2009. Cascading top-down effects of changing oceanic predator abundances. J. Anim. Ecol. 78(4), 699–714.

Bearhop, S., Adams, C. E., Waldron, S., Fuller, R. A. & Macleod, H. 2004. Determining trophic niche width: A novel approach using stable isotope analysis. J. Anim. Ecol. 73(5), 1007–1012.

Belleggia, M., Figueroa, D. E., Sánchez, F. & Bremec, C. 2012. Long-term changes in the spiny dogfish (*Squalus acanthias*) trophic role in the southwestern Atlantic. Hydrobiologia 684(1), 57–67.

Bethea, D. M., Bucker, J. A. & Carlson, J. K. 2004. Foraging ecology of the early life stages of four sympatric shark species. Mar. Ecol. Prog. Ser. 268, 245–264.

Bethea, D. M., Carlson, J. K., Buckel, J. A. & Satterwhite, M. 2006. Ontogenetic and site-related trends in the diet of the Atlantic sharpnose shark *Rhizoprionodon terraenovae* from the northeast Gulf of Mexico. Bull. Mar. Sci. 78(2), 287–307.

Bethea, D. M., Hale, L., Carlson, J. K., Cortés, E., Manire, C. A. & Gelsleichter, J. 2007. Geographic and ontogenetic variation in the diet and daily ration of the bonnethead shark, *Sphyrna tiburo*, from the eastern Gulf of Mexico. Mar. Biol. 152(5), 1009–1020.

Bigman, J. 2013. Trophic ecology of north pacific spiny dogfish (*Squalus suckleyi*) off central California waters. MSc book, California State University, United States.

Bizzarro, J. J., Broms, K. M., Logsdon, M. G., Ebert, D. A., Yoklavich, M. M., Kuhnz, L. A. & Summers, A. P. 2014. Spatial segregation in Eastern North pacific skate assemblages.

PLoS One 9(10), e109907.

Bizzarro, J. J., Robinson, H. J., Rinewalt, C. S. & Ebert, D. A. 2007. Comparative feeding ecology of four sympatric skate species off central California, USA. Environ. Biol. Fishes. 80, 197–220.

Blanco-Parra, M., Galván-Magaña, F., Márquez-Farías, J. F. & Niño-Torres, C. A. 2012. Feeding ecology and trophic level of the banded guitarfish, *Zapteryx exasperata*, inferred from stable isotopes and stomach contents analysis. Environ. Biol. Fishes. 95(1), 65–77.

Bonnici, L., Bonello, J. J. & Schembri, P. J. 2018. Diet and trophic level of the longnose spurdog, *Squalus blainville* (Risso, 1826) in the 25-nautical mile Fisheries Management Zone around the Maltese Islands. Reg. Stud. Mar. Sci. 19, 33–42.

Bornatowski, H., Wosnick, N., do Carmo, W. P. D., Corrêa, M. F. M. & Abilhoa, V. 2014. Feeding comparisons of four batoids (Elasmobranchii) in coastal waters of southern Brazil. J. Mar. Biolog. Assoc. U. K. 94(7), 1491–1499.

Braccini, J. M. 2008. Feeding ecology of two high-order predators from south-eastern Australia: The coastal broadnose and the deepwater sharpnose sevengill sharks. Mar. Ecol. Prog. Ser. 371, 273–284.

Braccini, J. M., Gillanders, B. M. & Walker, T. I. 2005. Sources of variation in the feeding ecology of the piked spurdog (*Squalus megalops*): Implications for inferring predator-prey interactions from overall dietary composition. ICES J. Mar. Sci. 62(6), 1076–1094.

Braccini, J. M., Gillanders, B. M. & Walker, T. I. 2006a. Notes on the population structure of the piked spurdog (*Squalus megalops*) in southeastern Australia. Cienc. Mar. 32(4), 705–712.

Braccini, J. M., Gillanders, B. M. & Walker, T. I. 2006b. Hierarchical approach to the assessment of fishing effects on non-target chondrichthyans: case study of *Squalus megalops* in southeastern Australia. Can. J. Fish. Aquat. Sci. 63(11), 2456–2466.

Braga, R. R., Bornatowski, H. & Vitule, J. R. S. 2012. Feeding ecology of fishes: An overview of worldwide publications. Rev. Fish Biol. Fish. 22(4), 915–929.

Branch, G. & Branch, M. 2018. Living shores Interacting with Southern Africa's marine ecosystems. 2nd ed. Penguin Random House, Cape Town. 336 pp.

Brett, C. & Walker, S. 2002. Predators and predation in Paleozoic marine environments. Paleontol. Soc. Pap. 8, 93–118.

Brickle, P., Laptikhovsky, V., Pompert, J. & Bishop, A. 2003. Ontogenetic changes in the feeding habits and dietary overlap between three abundant rajid species on the Falkland Islands' shelf. J. Mar. Biol. Assoc. U. K. 83(5), 1119–1125.

Broennimann, O., Fitzpatrick, M. C., Pearman, P. B., Petitpierre, B., Pellissier, L., Yoccoz, N. G., Thuiller, W., Fortin, M. J., Randin, C., Zimmermann, N. E., Graham, C. H. & Guisan, A. 2012. Measuring ecological niche overlap from occurrence and spatial environmental data. Glob. Ecol. Biogeogr. 21(4), 481–497.

Brown, J. 2004. Towards a metabolic theory of ecology. Ecology 85, 1771–1789.

Brown, S. C., Bizzarro, J. J., Cailliet, G. M. & Ebert, D. A. 2012. Breaking with tradition: Redefining measures for diet description with a case study of the Aleutian skate *Bathyraja aleutica* (Gilbert 1896). Environ. Biol. Fishes 95(1), 3–20.

Bush, A. 2003. Diet and diel feeding periodicity of juvenile scalloped hammerhead sharks, S*phyrna lewini*, in Kāne'ohe Bay, Ō'ahu, Hawai'i. Environ. Biol. Fishes 67(1), 1–11.

Butler, P. & Taylor, E. 1975. The effect of progressive hypoxia on respiration in the dogfish (*Scyliorhinus canicula*) at different seasonal temperatures. J. Exp. Biol. 63(1), 117–130.

Cailliet, G., Musick, J., Simpfendorfer, C. & Stevens, J. 2005. Ecology and life history characteristics of chondrichthyan fish. In: SSC Shark Specialist Group IUCN, Sharks, rays and chimaeras: the status of the chondrichthyan fishes. Eds Fowler, S., Cavanagh, R., Camhi, M., Burgess, G., Cailliet, G., Fordham, S., Simpfendorfer, C., Musick, J. Cambridge Press, United Kingdom. pp.12-18.

Carlisle, A. B., Kim, S. L., Semmens, B. X., Madigan, D. J., Jorgensen, S. J., Perle, C. R., Anderson, S. D., Chapple, T. K., Kanive, P. E. & Block, B. A. 2012. Using stable isotope analysis to understand the migration and trophic ecology of northeastern Pacific white sharks (*Carcharodon carcharias*). PLoS One 7(2), e30492.

Carrassón, M., Stefanescu, C. & Cartes, J. E. 1992. Diets and bathymetric distributions of two bathyal sharks of the Catalan deep sea (western Mediterranean). Mar. Ecol. Prog. Ser. 82, 21–30.

Carrier, J., Musick, J. & Heithaus, M. 2004. Biology of Sharks and Their Relatives. 1st ed.

CRC Press, Boca Raton. 246 pp.

Carrier, J., Musick, J. & Heithaus, M. 2012. Biology of Sharks and Their Relatives. 2nd ed. Eds Carrier, J., Musick, J. & Heithaus, M. CRC Press, New York. 666 pp.

Carroll, R. L. 1988. Vertebrate Palaeontology and Evolution. Macmillan Publishers, New York. 698 pp.

Cavanagh, R. D. & Gibson, C. 2007. Overview of the conservation status of cartilaginous fishes (Chondrichthyans) in the Mediterranean Sea. IUCN, Spain. 42 pp.

Cavanagh, R. & Kyne, P. 2006. The conservation status of deep-sea chondrichthyan fishes. FAO Fish Proc. 3, 366-380.

Ceccarelli, D. M., Frisch, A. J., Graham, N. A. J., Ayling, A. M. & Beger, M. 2014. Habitat partitioning and vulnerability of sharks in the Great Barrier Reef Marine Park. Rev. Fish Biol. Fish. 24(1), 169–197.

Churchill, D. A., Heithaus, M. R., Vaudo, J. J., Grubbs, R. D., Gastrich, K. & Castro, J. I. 2015. Trophic interactions of common elasmobranchs in deep-sea communities of the Gulf of Mexico revealed through stable isotope and stomach content analysis. Deep Sea Res. Part II Top. Stud. Oceanogr. 115, 92–102.

Clarke, M. 1986. A handbook for the identification of cephalopod beaks. Clarendon Press, United Kingdom. 273 pp.

Clarke, M. W., Borges, L. & Officer, R. A. 2005. Comparisons of trawl and longline catches of deepwater elasmobranchs west and north of Ireland. J. Northwest Atl. Fish. Sci. 35, 429–442.

Cochrane, K. L., Augustyn, C. J., Cockcroft, A. C., David, J. H. M., Griffiths, M. H., Groeneveld, J. C., Lipi, M., Smale, M. J., Smith, C. D. & Tarr, R. J. Q. 2004. An ecosystem approach to fisheries in the southern Benguela context. In: Ecosystem approaches to fisheries in the southern Benguela. Eds Shannon, L. J., Cochrane, K. L. & Pillar, S. C. Afr. J. Mar. Sci. 26, 9–35

Cohen, J., Pimm, S., Yodzis, P. & Saldana, J. 1993. Body sizes of animal predators and animal prey in food webs. J. Anim. Ecol. 62, 67–78.

Collins, M. A., Bailey, D. M., Ruxton, G. D. & Priede, I. G. 2005. Trends in body size across an environmental gradient: A differential response in scavenging and non-scavenging demersal deep-sea fish. Proc. Royal Soc. B Biol. Sci. 272(1576), 2051–2057.

Colman, J. 1997. A review of the biology and ecology of the whale shark. J. Fish Biol. 51(6), 1219–1234.

Compagno, L. J. V. 1984. FAO species catalogue: Vol. 4, Part 2 - Sharks of the world. FAO Fish. Synop. 125, 295-305.

Compagno, L. 1990a. Shark exploitation and conservation. In: Elasmobranchs as Living Resources: Advances in the Biology, Ecology, Systematics, and Status of the Fisheries. Eds Pratt, Jr, H. L., Gruber, S. H. & Taniuchi, T. NOAA Tech. Rep. 90: 391-414.

Compagno, L. J. 1990b. Alternative life-history styles of cartilaginous fishes in time and space. Environ. Biol. Fishes 28, 33–75.

Compagno, L. 1999. Checklist of Living Elasmobranchs. In: Sharks, skates, and rays: The biology of elasmobranch fishes. Eds Hamlett, W. Johns Hopkins University Press, Baltimore. pp. 471–498.

Compagno, L., Dando, M. & Fowler, S. 2005. A Field Guide to the Sharks of the World. Harper Collins Publishing Ltd. London. 368 pp.

Compagno, L. J. V., Ebert, D. A. & Cowley, P. D. 1991. Distribution of offshore demersal cartilaginous fish (Class Chondrichthyes) off the west coast of southern Africa, with notes on their systematics. S. Afr. J. Mar. Sci. 11(1), 43–139.

Connell, J. 1983. On the prevalence and relative importance of interspecific competition: evidence of field experiments. Am. Nat. 122(5), 661–696.

Cooper, D. 1968. The significance level in multiple tests made simultaneously. Heredity 23(4), 614–617.

Cortés, E. 1997. A critical review of methods of studying fish feeding based on analysis of stomach contents: application to elasmobranch fishes. Can. J. Fish. Aquat. Sci. 54(3), 726–738.

Cortés, E. 1998. Methods of studying fish feeding: reply. Can. J. Fish. Aquat. Sci. 55(12), 2708.

Cortés, E. 1999. Standardized diet compositions and trophic levels of sharks. ICES J. Mar. Sci. 56(5), 707–717.

Cortés, E. & Gruber, S. H. 1990. Diet, feeding habits and estimates of daily ration of young lemon sharks, *Negaprion brevirostris* (Poey). Copeia 1990(1), 204–218.

Cotton, C. F., Grubbs, R. D., Daly-Engel, T. S., Lynch, P. D. & Musick, J. A. 2011. Age, growth and reproduction of a common deep-water shark, shortspine spurdog (*Squalus cf. mitsukurii*), from Hawaiian waters. Mar. Freshw. Res. 62(7), 811–822.

Courtney, D. L. & Foy, R. 2012. Pacific sleeper shark *Somniosus pacificus* trophic ecology in the eastern North Pacific Ocean inferred from nitrogen and carbon stable-isotope ratios and diet. J. Fish Biol. 80(5), 1508–1545.

Currie, J. C., Atkinson, L. J., Sink, K. J. & Attwood, C. G. 2020. Long-Term Change of Demersal Fish Assemblages on the Inshore Agulhas Bank Between 1904 and 2015. Front. Mar. Sci. 7, 355.

Davenport, S. R. & Bax, N. J. 2002. A trophic study of a marine ecosystem off southeastern Australia using stable isotopes of carbon and nitrogen. Can. J. Fish. Aquat. Sci. 59(3), 514–530.

DeAngelis, B. M., McCandless, C. T., Kohler, N. E., Recksiek, C. W. & Skomal, G. B. 2008. First characterization of shark nursery habitat in the United States Virgin Islands: Evidence of habitat partitioning by two shark species. Mar. Ecol. Prog. Ser. 358, 257–271.

DeNiro, M. J. & Epstein, S. 1978. Influence of diet on the distribution of carbon isotopes in animals. Geochim. Cosmochim. Acta 42(5), 495–506.

de Sousa Rangel, B., Hussey, N. E., Gomes, A. D., Rodrigues, A., Martinelli, L. A. & Moreira, R. G. 2019. Resource partitioning between two young-of-year cownose rays *Rhinoptera bonasus* and *R. brasiliensis* within a communal nursery inferred by trophic biomarkers. J. Fish Biol. 94(5), 781–788.

Diaz, C. B. 2014. Biology, Taxonomy and distribution of south-east Pacific cartilaginous fishes. PhD book, University of Queensland, Australia.

Diaz, R. J. & Rosenberg, R. 2008. Spreading dead zones and consequences for marine ecosystems. Science 321(5891), 926–929.

Dicken, M. L., Hussey, N. E., Christiansen, H. M., Smale, M. J., Nkabi, N., Cliff, G. &

Wintner, S. P. 2017. Diet and trophic ecology of the tiger shark (*Galeocerdo cuvier*) from South African waters. PLoS One 12(6), e0177897.

Dippenaar, S. & Molele, R. 2015. Siphonostomatoid copepods infecting *Squalus acutipinnis* Regan, 1908 off South Africa. Afr. J. Mar. Sci. 37(4), 605–608.

Domi, N., Bouquegneau, J. M. & Das, K. 2005. Feeding ecology of five commercial shark species of the Celtic Sea through stable isotope and trace metal analysis. Mar. Environ. Res. 60(5), 551–569.

Dulvy, N. & Forrest, R. 2010. Life histories, population dynamics, and extinction risks in chondrichthyans. In: Sharks and their relatives II: biodiversity, adaptive physiology, and conservation. Eds Carrier, J., Musick, J., Heithaus, M. CRC Press, Boca Raton. pp. 639–679.

Dulvy, N. K., Fowler, S. L., Musick, J. A., Cavanagh, R. D., Kyne, P. M., Harrison, L. R., Carlson, J. K., Davidson, L. N. k, Fordham, S. V., Francis, M. P., Pollock, C. M., Simpfendorfer, C. A., Burgess, G. H., Carpenter, K. E., Compagno, L. J. v, Ebert, D. A., Gibson, C., Heupel, M. R., Livingstone, S. R., Sanciangco, J. C., Stevens, J. D., Valenti, S. & White, W. T. 2014. Extinction risk and conservation of the world's sharks and rays. Elife 3, e00590.

Dulvy, N., Metcalfe, J., Glanville, J., Pawson, M. & JD, R. 2000. Fishery stability, local extinctions, and shifts in community structure in skates. Conserv. Biol. 14(1), 283–293.

Dunn, M. R., Stevens, D. W., Forman, J. S. & Connell, A. 2013. Trophic interactions and distribution of some Squaliforme sharks, including new diet descriptions for *Deania calcea* and *Squalus acanthias*. PLoS One 8(3), e59938.

Ebert, D. A. 2013. FAO Species Catalogue for Fishery Purposes (6): Deep-Sea Cartilaginous Fishes of the Indian Ocean. FAO, Rome. 129 pp.

Ebert, D. A. 2015. FAO Species Catalogue for Fishery Purposes (9): Deep-sea cartilaginous fishes of the Southeastern Atlantic Ocean. FAO, Rome. 251 pp.

Ebert, D., Compagno, L. & Cowley, P. 1992. A preliminary investigation of the feeding ecology of squaloid sharks off the west coast of southern Africa. S. Afr. J. Mar. Sci. 12(1), 601–609.

Ebert, D. A., Haas, D. L. & De Carvalho, M. R. 2015. *Tetronarce cowleyi*, sp. nov., a new species of electric ray from southern Africa (Chondrichthyes: Torpediniformes: Torpedinidae). Zootaxa 3936(2), 237–250.

Ebert, D. A. & Mostarda, E. 2013. Identification Guide to the Deep – Sea Cartilaginous Fishes of the Indian Ocean. FishFinder Programme. FAO, Rome. 76 pp.

Ebert, D. A. & Mostarda, E. 2016. Identification guide to the deep-sea cartilaginous fishes of the Southeastern Pacific Ocean. FishFinder Programme. FAO, Rome. 53 pp.

Ebert, D. A. & Stehman, M. F. 2013. FAO Species Catalogue for Fishery Purposes (7): Sharks, batoids, and chimaeras of the North Atlantic. FAO, Rome. 523 pp.

Ebert, D. A., White, W. T., Goldman, K. J., Compagno, L. J. V., T.S, D. E. & Ward., R. D. 2010. Resurrection and rediscription of *Squalus suckleyi* (Girard,1854) from the North Pacific, with comments on the *Squalus acanthias* subgroup (Squaliformes:Squaloidae). Zootaxa 2612(1), 22–40.

Eschmeyer, W. 2014. Catalog of fishes: genera, species, references. http://researcharchive.calacademy.org/research/ichthyology/catalog/fishcatmain.asp.

Espinoza, M., Clarke, T. M., Villalobos-Rojas, F. & Wehrtmann, I. S. 2012. Ontogenetic dietary shifts and feeding ecology of the rasptail skate *Raja velezi* and the brown smoothhound shark *Mustelus henlei* along the Pacific coast of Costa Rica, Central America. J. Fish Biol. 81(5), 1578–1595.

Espinoza, M., Munroe, S. E. M., Clarke, T. M., Fisk, A. T. & Wehrtmann, I. S. 2015. Feeding ecology of common demersal elasmobranch species in the Pacific coast of Costa Rica inferred from stable isotope and stomach content analyses. J. Exp. Mar. Biol. Ecol. 470, 12–25.

Estrada, J. A., Rice, A. N., Lutcavage, M. E. & Skomal, G. B. 2003. Predicting trophic position in sharks of the north-west Atlantic Ocean using stable isotope analysis. J. Mar. Biol. Assoc. U. K. 83(6), 1347–1350.

Estrada, J. A., Rice, A. N., Natanson, L. J. & Skomal, G. B. 2006. Use of isotopic analysis of vertebrae in reconstructiong ontogenetic feeding ecology in white sharks. Ecology 87(4), 829–834 .

Fairclough, D. V., Clarke, K. R., Valesini, F. J. & Potter, I. C. 2008. Habitat partitioning by five congeneric and abundant *Choerodon* species (Labridae) in a large subtropical marine embayment. Estuar. Coast. Shelf Sci. 77(3), 446–456.

Fanelli, E., Rey, J., Torres, P. & Gil De Sola, L. 2009. Feeding habits of blackmouth catshark *Galeus melastomus* Rafinesque, 1810 and velvet belly lantern shark *Etmopterus spinax* (Linnaeus, 1758) in the western Mediterranean. J. Appl. Ichthyol. 25, 83–93.

Ferretti, F., Myers, R. A., Serena, F. & Lotze, H. K. 2008. Loss of large predatory sharks from the Mediterranean Sea. Conserv. Biol. 22(4), 952–964.

Ferretti, F., Osio, G. C., Jenkins, C. J., Rosenberg, A. A. & Lotze, H. K. 2013. Long-term change in a meso-predator community in response to prolonged and heterogeneous human impact. Sci. Rep. 3, 1–11.

Ferretti, F., Worm, B., Britten, G. L., Heithaus, M. R. & Lotze, H. K. 2010. Patterns and ecosystem consequences of shark declines in the ocean. Ecol. Lett. 13(8), 1055–1071.

Ferry, L. & Cailliet, G. 1996. Sample size and data analysis: are we characterizing and comparing diet properly? In: GUTSHOP 96 Proceedings of the symposium on the Feeding Ecology and Nutrition in Fish. Eds MacKinlay, D., Shearer, K. San Francisco State University, United States. 71-80 pp.

Figueirêdo, S. T. V. de. 2011. Taxonomic and morphological review of the genus *Squalus* Linnaeus, 1758 of the Western South Atlantic Ocean (Chondrichthyes: Squaliformes: Squaloidae). PhD book, University of Sao Paulo, Brazil.

Fischer, A. F., Veras, D. P., Hazin, F. H. V., Broadhurst, M. K., Burgess, G. H. & Oliveira, P. G. V. 2006. Maturation of *Squalus mitsukurii* and *Cirrhigaleus asper* (Squaloidae, Squaliformes) in the southwestern equatorial Atlantic Ocean. J. Appl. Ichthyol. 22(6), 495–501.

Fisk, A., Tittlemier, S., Pranschke, J. & Norstrom, R. 2002. Using anthropogenic contaminants and stable isotopes to assess the feeding ecology of the elusive Greenland shark (*Somniosus microcephalus*). Ecology 83(8), 2162–2172.

Flammang, B. E., Ebert, D. A. & Cailliet, G. M. 2011. Intraspecific and Interspecific Spatial Distribution of Three Eastern North Pacific Catshark Species and Their Egg Cases (Chondrichthyes: Scyliorhinidae). Breviora 525 (1), 1–18.

Fowler, S., Cavanagh, R., Camhi, M., Burgess, G., Cailliet, G., SV Simpfendorfer, C. F. & Musick, J. 2005. Sharks, Rays and Chimaeras: The Status of the Chondrichthyan Fishes: Status Survey. IUCN, Cambridge, United Kingdom. 462 pp.

France, R. 1995. Carbon-13 enrichment in benthic compared to planktonic algae: foodweb implications. Mar. Ecol. Prog. Ser. 124, 307–312.

France, R. L. & Peters, R. H. 1997. Ecosystem differences in the trophic enrichment of 13 C in aquatic food webs. Can. J. Fish. Aquat. 54(6), 1255–1258.

Fricke, R., Eschmeyer, W. & van der Laan, R. 2020. Eschmeyer's Catalog of Fishes: genera, species, references. http://researcharchive.calacademy.org/research/ichthyology/catalog/fishcatmain.asp.

Frisk, M. 2010. Life history strategies of batoids. In: Sharks and Their Relatives II: Biodiversity, Adaptive Physiology, and Conservation. 2nd ed. Eds Carrier, J. C., Musick, J. A. & Heithaus, M. R. CRC Press, Florida. pp. 283–316.

Frisk, M., Miller, T. & Fogarty MJ. 2001. Estimation and analysis of biological parameters in elasmobranch fishes: a comparative life history study. Can. J. Fish. Aquat. Sci. 58(5), 969–981.

Froese, R. & Pauly, D. 2019. Squaloidae. Fishbase. https://www.fishbase.de/summary/FamilySummary.php?ID=13.

Fry, B. 1996. $^{13}C/^{12}C$ fractionation by marine diatoms. Mar. Ecol. Prog. Ser. 134, 283–294.

Gannes, L. Z., Brien, D. M. O. & Martinez del Rio, C. 1997. Stable isotopes in animal ecology : assumptions, caveats, and a call for more laboratory experiments. Ecology 78(4), 1271–1276.

Garrison, L. P. & Link, J. S. 2000. Dietary guild structure of the fish community in the Northeast United States continental shelf ecosystem. Mar. Ecol. Prog. Ser. 202, 231–240.

Gelsleichter, J., Musick, J. A. & Nichols, S. 1999. Food habits of the smooth dogfish, *Mustelus canis*, dusky shark, *Carcharhinus obscurus*, Atlantic sharpnose shark, *Rhizoprionodon terraenovae*, and the sand tiger, *Carcharias taurus*, from the northwest Atlantic Ocean. Environ. Biol. Fishes 54(2), 205–217.

Graham, K. J. 2005. Distribution, population structure and biological aspects of *Squalus* spp. (Chondrichthyes : Squaliformes) from New South Whales and adjacent Australian waters. Mar. Freshw. Res. 56(4), 405–416.

Graham, K. J. & Daley, R. K. 2011. Distribution, reproduction and population structure of three gulper sharks (*Centrophorus*, Centrophoridae) in south-east Australian waters.

Mar. Freshw. Res. 62(6), 583–595.

Grogan, E., Lund, R. & Greenfest-Allen, E. 2012. The Origin and Relationships of Early Chondrichthyans. In: Biology of Sharks and Their Relatives. Eds Carrier, J., Musick, J., Heithaus, M. 2nd ed. CRC Press, London. pp. 3-30.

Guisan, A. & Thuiller, W. 2005. Predicting species distribution: Offering more than simple habitat models. Ecol. Lett. 8(9), 993–1009.

Hamlett, W. C. 2005. Reproductive biology and phylogeny of Chondrichthyes: Sharks, batoids and chimaeras. 1st ed. Science Publishers, New Hampshire. 576 pp.

Hanchet, S. 1988. Reproductive biology of *Squalus acanthias* from the east coast, South Island, New Zealand. N. Z. J. Mar. Freshwater Res. 55(4), 537–549.

Hanchet, S. 1991. Diet of spiny dogfish, *Squalus acanthias* Linnaeus, on the east coast, South Island, New Zealand. J. Fish Biol. 39(3), 313–323.

Hastings, P., Walker, H. J. & Galland, G. R. 2015. Fishes: A Guide to Their Diversity. University of California Press, California. 352 pp.

Heemstra, P. & Heemstra, E. 2004. Coastal Fishes of Southern Africa. NISC (PTY) LTD, Grahamstown. 488 pp.

Heinicke, M., Naylor, G. & Hedges, S. 2009. Catilaginous Fishes. In: The Timetree of Life. Eds Hedges, S., Kumar, S. Oxford University Press, New York. pp. 320-327.

Heithaus, M. R. 2001. Predator–prey and competitive interactions between sharks (order Selachii) and dolphins (suborder Odontoceti): a review. J. Zool. 253(1), 53–68.

Heithaus, M. R., Dill, L. M., Marshall, G. J. & Buhleier, B. 2002. Habitat use and foraging behavior of tiger sharks (*Galeocerdo cuvier*) in a seagrass ecosystem. Mar. Biol. 140(2), 237–248.

Heithaus, M. R., Frid, A., Wirsing, A. J. & Worm, B. 2008. Predicting ecological consequences of marine top predator declines. Trends Ecol. Evol. 23(4), 202–210.

Heithaus, M., Vaudo, J., Kreicker, S., Layman, C., Krutzen, M., Burkholder, D., Bessey, C., Sarabia, R., Cameron, K., Wirsing, A., Thomson, J. & Dunphy-Daly, M. 2013. Apparent resource partitioning and trophic structure of large-bodied marine predators in a relatively pristine seagrass ecosystem. Mar. Ecol. Prog. Ser. 481, 225–237.

Helfman, G., Collette, B., Facey, D. & Bowen, B. 2009. The diversity of fishes: biology, evolution, and ecology. 2nd ed. John Wiley & Sons, Oxford. 736 pp.

Heupel, M. R., Knip, D. M., Simpfendorfer, C. A. & Dulvy, N. K. 2014. Sizing up the ecological role of sharks as predators. Mar. Ecol. Prog. Ser. 495, 291–298.

Heyns, E. R., Bernard, A. T. F., Richoux, N. B. & Götz, A. 2016. Depth-related distribution patterns of subtidal macrobenthos in a well-established marine protected area. Mar. Biol. 163(2), 39.

Hill, J. M. & McQuaid, C. D. 2008. δ^{13}C and δ^{15}N biogeographic trends in rocky intertidal communities along the coast of South Africa: Evidence of strong environmental signatures. Estuar. Coast. Shelf Sci. 80(2), 261–268.

Hill, J. M., McQuaid, C. D. & Kaehler, S. 2006. Biogeographic and nearshore-offshore trends in isotope ratios of intertidal mussels and their food sources around the coast of southern Africa. Mar. Ecol. Prog. Ser. 318, 63–73.

Hobson, K. 1999. Tracing origins and migration of wildlife using stable isotopes: a review. Oecologia 120(3), 314–326.

Hobson, K., Alisauskas, R. & Clark, R. 1993. Stable nitrogen enrichment in avian tissues due to fasting and nutritional stress - implications for isotopic analyses of diet. Condor 95(2), 388–394.

Hobson, K. A., Schell, D. M., Renouf, D. & Noseworthy, E. 1996. Stable carbon and nitrogen isotopic fractionation between diet and tissues of captive seals: implications for dietary reconstructions involving marine mammals. Can. J. Fish. Aquat. Sci. 53(3), 528–533.

Hoffman, J. & Sutton, T. 2010. Lipid correction for carbon stable isotope analysis of deep-sea fishes. Deep Sea Res. Part I Oceanogr. Res. Pap. 57(8), 956–964.

Hoffmayer, E. R. & Parsons, G. R. 2003. Food habits of three shark species from the Mississippi Sound in the northern Gulf of Mexico. Southeast. Nat. 2(2), 271–280.

Hopkins, T. E. & Cech, J. J. 2003. The influence of environmental variables on the distribution and abundance of three elasmobranchs in Tomales Bay, California. Environ. Biol. Fishes 66(3), 279–291.

Huckembeck, S., Loebmann, D., Albertoni, E. F., Hefler, S. M., Oliveira, M. C. L. M. & Garcia, A. M. 2014. Feeding ecology and basal food sources that sustain the Paradoxal frog *Pseudis minuta*: A multiple approach combining stomach content, prey availability, and stable isotopes. Hydrobiologia 740, 253–264.

Hussey, N. E., Dudley, S. F. J., Mccarthy, I. D., Cliff, G. & Fisk, A. T. 2011. Stable isotope profiles of large marine predators : Viable indicators of trophic position , diet , and movement in sharks ? Can. J. Fish. Aquat. Sci. 68(12), 2029–2045.

Hussey, N. E., Macneil, M. A., Mcmeans, B. C., Olin, J. A., Dudley, S. F. J., Cliff, G., Wintner, S. P., Fennessy, S. T. & Fisk, A. T. 2014. Rescaling the trophic structure of marine food webs. Ecol. Lett. 17(2), 239–250.

Hussey, N. E., MacNeil, M. A., Olin, J. A., McMeans, B. C., Kinney, M. J., Chapman, D. D. & Fisk, A. T. 2012. Stable isotopes and elasmobranchs: Tissue types, methods, applications and assumptions. J. Fish Biol. 80(5), 1449–1484.

Hutchings, K. & Lamberth, S. 2002. Bycatch in the gillnet and beach-seine fisheries in the Western Cape, South Africa, with implications for management. S. Afr. J. Mar. Sci. 24(1), 227–241.

Hutchings, L., van der Lingen, C. D., Shannon, L. J., Crawford, R. J. M., Verheye, H. M. S., Bartholomae, C. H., van der Plas, A. K., Louw, D., Kreiner, A., Ostrowski, M., Fidel, Q., Barlow, R. G., Lamont, T., Coetzee, J., Shillington, F., Veitch, J., Currie, J. C. & Monteiro, P. M. S. 2009. The Benguela Current: An ecosystem of four components. Prog. Oceanogr. 83(1-4), 15–32.

Hutchings, J., Minto, C., Ricard, D., Baum, J. & Jensen, O. 2010. Trends in the abundance of marine fishes. Can. J. Fish. Aquat. Sci. 67(8), 1205–1210.

Hyndes, G., Platell, M., Potter, I. & Lenanton, R. 1999. Does the composition of the demersal fish assemblages in temperate coastal waters change with depth and undergo consistent seasonal changes? Mar. Biol. 134(2), 335–352.

Hyslop, E. J. 1980. Stomach contents analysis-a review of methods and their application. J. Fish Biol 17(4), 441-429.

Jacoby, D. M. P., Croft, D. P. & Sims, D. W. 2012. Social behaviour in sharks and rays: Analysis, patterns and implications for conservation. Fish Fish. 13(4), 399–417.

Jakobsdóttir, K. B. 2001. Biological aspects of two deep-water squaloid sharks:

Centroscyllium fabricii (Reinhardt, 1825) and *Etmopterus princeps* (Collett, 1904) in Icelandic waters. Fish. Res. 51(2-3), 247–265.

Hansson, S., Hobbie, J. E., Elmgren, R., Larsson, U., Fry, B. & Johansson, S. 1997. The stable nitrogen isotope ratio as a marker of food-web interactions and fishmigration. Ecology 78(7), 2249-2257.

Jones, M. E. & Barmuta, L. A. 1998. Diet overlap and relative abundance of sympatric dasyurid carnivores: A hypothesis of competition. J. Anim. Ecol. 67(3), 410–421.

Jordan, D. & Snyder, J. 1903. On a collection of fishes made by Mr. Alan Owston in the deep waters of Japan. Smithsonian Miscellaneous Collections. Washington.

Kadila, H. 2019. Trophic relationships of shallow water Cape hake (*Merluccius capensis*) and Cape horse mackerel (*Trachurus capensis*) in the northern Benguela Ecosystem. MSc. book, University of Namibia, Namibia.

Katzenberg, M. A. 2008. Stable isotope analysis: a tool for studying past diet, demography, and life history. Biol. Anthropol. Hum. Skelet. 2, 413–441.

Kazunari, Y. A. N. O. 1991. Catch distribution, stomach contents and size at maturity of two Squaloid sharks, *Deania calceus* and *D. crepidalbus*, from the Southeast Atlantic off Namibia. Bull. Japan Soc. Fish. Ocean. 55(3), 189–196.

Kazunari, Y. A. N. O., & Tanaka, S. 1988. Size at maturity, reproductive cycle, fecundity, and depth segregation of the deep sea squaloid sharks *Centroscymnus owstoni* and *C. coelolepis* in Suruga Bay, Japan. Nippon Suisan Gakkai Shi 54(2), 167–174.

Ketchen, K. 1986. The spiny dogfish (*Squalus acanthias*) in the northeast Pacific and a history of its utilization. Can. Spec. Publ. Fish. Aquat. Sci. 88, 68–88.

Kim, S. L. & Koch, P. L. 2012. Methods to collect, preserve, and prepare elasmobranch tissues for stable isotope analysis. Environ. Biol. Fishes 95(1), 53–63.

Kim, S. L., del Rio, C. M., Casper, D. & Koch, P. L. 2012. Isotopic incorporation rates for shark tissues from a long-term captive feeding study. J. Exp. Biol. 215(14), 2495–2500.

Kinney, M. J., Hussey, N. E., Fisk, A. T., Tobin, A. J. & Simpfendorfer, C. A. 2011. Communal or competitive? Stable isotope analysis provides evidence of resource partitioning within a communal shark nursery. Mar. Ecol. Prog. Ser. 439, 263–276.

Kiszka, J. J., Charlot, K., Hussey, N. E., Heithaus, M. R., Simon-Bouhet, B., Humber, F., Caurant, F. & Bustamante, P. 2014. Trophic ecology of common elasmobranchs exploited by artisanal shark fisheries off south-western Madagascar. Aquat. Biol. 23(1), 29–38.

Klimley, A. P. 1987. The determinants of sexual segregation in the scalloped hammerhead shark, *Sphyrna lewini*. Environ. Biol. Fishes 18(1), 27–40.

Klimley, A. 2013. The Biology of Sharks and Rays. University of Chicago Press, Chicago. 488 pp.

Knip, D. M., Heupel, M. R. & Simpfendorfer, C. A. 2010. Sharks in nearshore environments: models, importance, and consequences. Mar. Ecol. Prog. Ser. 402, 1–11.

Knip, D. M., Heupel, M. R. & Simpfendorfer, C. A. 2012. Habitat use and spatial segregation of adult spottail sharks *Carcharhinus sorrah* in tropical nearshore waters. J. Fish Biol. 80(4), 767–784.

Knip, D. M., Heupel, M. R., Simpfendorfer, C. A., Tobin, A. J. & Moloney, J. 2011. Ontogenetic shifts in movement and habitat use of juvenile pigeye sharks *Carcharhinus amboinensis* in a tropical nearshore region. Mar. Ecol. Prog. Ser. 425, 233–246.

Krebs, C. J. 2001. Ecology: The Experimental Analysis of Distribution and Abundance. 5th ed. Benjamin Cummings, California. 695 pp.

Kyne, P. M., Compagno, L. J. V, Stead, J., Jackson, M. V. & Bennett, M. B. 2011. Distribution, habitat and biology of a rare and threatened eastern Australian endemic shark: Colclough's shark, *Brachaelurus colcloughi* Ogilby, 1908. Mar. Freshw. Res. 62(6), 540–547.

Kyne, P.M. & Heupel, M.R. 2015. *Squaliolus aliae. The IUCN Red List of Threatened Species* 2015: e.T41858A68643995. https://dx.doi.org/10.2305/IUCN.UK.2015-RLTS.T41858A68643995.en.

Laptikhovsky, V. V., Arkhipkin, A. I. & Henderson, A. C. 2001. Feeding habits and dietary overlap in spiny dogfish *Squalus acanthias* (Squaloidae) and narrowmouth catshark *Schroederichthys bivius* (Scyliorhinidae). J. Mar. Biol. Assoc. U. K. 81, 1015–1018.

Last, P. & Stevens, J. 2009. Sharks and Rays of Australia. 2nd ed. CSIRO Publishing, Collingwood. 644 pp.

Last, P., White, W. & Pogonoski, J. 2007. Descriptions of new dogfishes of the genus
Squalus (Squaloidea: Squaloidae). CSIRO Marine and Atmospheric Research, Hobart.
130 pp.

Leslie, R., Glazer, J. & Fairweather, T. 2013. Discussion paper on the calibration of Africana
against selected commercial vessels. Maram IWS.
http://www.maram.uct.ac.za/sites/default/files/image_tool/images/302/workshop/2012/
MARAM_IWS_NOV12_SURV_HK_P1.pdf.

Linke, T. E., Platell, M. E. & Potter, I. C. 2001. Factors influencing the partitioning of food
resources among six fish species in a large embayment with juxtaposing bare sand and
seagrass habitats. J. Exp. Mar. Bio. Ecol. 266(2), 193–217.

Lisney, T. 2010. A review of the sensory biology of chimaeroid fishes (Chondrichthyes;
Holocephali). Rev. Fish Biol. Fish. 20(4), 571–590.

Logan, J. M. & Lutcavage, M. E. 2010. Stable isotope dynamics in elasmobranch fishes.
Hydrobiologia 644(1), 231–244.

López-García, J., Navia, A. F., Mejía-Falla, P. A. & Rubio, E. A. 2012. Feeding habits and
trophic ecology of *Dasyatis longa* (Elasmobranchii: Myliobatiformes): sexual, temporal
and ontogenetic effects. J. Fish Biol. 80(5), 1563–1579.

Lopez, S., Barria, P. & Melendez, R. 2012. Feeding and trophic relationships of two highly
migratory sharks in the eastern south Pacific Ocean. Pan-Am. J. Aquat. Sci. 7(1), 50–56.

Lowe, C. G., Wetherbee, B. M., Crow, G. L. & Tester, A. L. 1996. Ontogenetic dietary shifts
and feeding behavior of the tiger shark, *Galeocerdo cuvier*, in Hawaiian waters.
Environ. Biol. Fishes 47(2), 203–211.

Lucifora, L. O., García, V. B., Menni, R. C. & Escalante, A. H. 2006. Food habits,
selectivity, and foraging modes of the school shark *Galeorhinus galeus*. Mar. Ecol.
Prog. Ser. 315, 259–270.

Lucifora, L. O., Valero, J. L., Bremec, C. S. & Lasta, M. L. 2000. Feeding habits and prey
selection by the skate *Dipturus chilensis* (Elasmobranchii: Rajidae) from the south-
western Atlantic. J. Mar. Biol. Assoc. U. K. 80(5), 953–954.

Lucifora, L., Valero, J. & García, V. 1999. Length at maturity of the greeneye spurdog shark,
Squalus mitsukurii (Elasmobranchii: Squaloidae), from the SW Atlantic, with
comparisons with other regions. Mar. Freshw. Res. 50(7), 629–632.

Lund, R. & Grogan, E. 1997. Relationships of the Chimaeriformes and the basal radiation of the Chondrichthyes. Rev. Fish Biol. Fish. 7(1), 65–123.

Lutjeharms, J. R. E. 2006. The Agulhas Current. 5th ed. Springer International Publishing, Berlin. 329 pp.

Macdonald, J. S. & Green, R. H. 1983. Redundancy of variables used to describe importance of prey species in fish diets. Can. J. Fish. Aquat. Sci. 40(5), 635–637.

Macleod, C. D. 2005. Niche Partitioning, distribution and competition in North Atlantic Beaked Whales. PhD book, University of Aberdeen, Scotland.

MacNeil, M. A., Skomal, G. B. & Fisk, A. T. 2005. Stable isotopes from multiple tissues reveal diet switching in sharks. Mar. Ecol. Prog. Ser. 302, 199–206.

Macpherson, E. & Duarte, C. 1991. Bathymetric trends in demersal fish size: is there a general relationship? Mar. Ecol. Prog. Ser. 71, 103–112.

Malpica-Cruz, L., Herzka, S. Z., Sosa-Nishizaki, O., Lazo, J. P. & Trudel, M. 2012. Tissue-specific isotope trophic discrimination factors and turnover rates in a marine elasmobranch: empirical and modeling results. Can. J. Fish. Aquat. Sci. 69(3), 551–564.

Marshall, A. D., Kyne, P. M. & Bennett, M. B. 2008. Comparing the diet of two sympatric urolophid elasmobranchs (*Trygonoptera testacea* Muller & Henle and *Urolophus kapalensis* Yearsley & Last): evidence of ontogenetic shifts and possible resource partitioning. J. Fish Biol. 72(4), 883–898.

Matich, P. 2014. Environmental and Individual Factors Shaping the Habitat Use and Trophic Interactions of Juvenile Bull Sharks (*Carcharhinus leucas*) in a Subtropical Estuary. PhD book, Florida International University, United States.

Matich, P. & Heithaus, M. R. 2014. Multi-tissue stable isotope analysis and acoustic telemetry reveal seasonal variability in the trophic interactions of juvenile bull sharks in a coastal estuary. J. Anim. Ecol. 83, 199–213.

Matich, P., Heithaus, M. R. & Layman, C. A. 2011. Contrasting patterns of individual specialization and trophic coupling in two marine apex predators. J. Anim. Ecol. 80(1), 294–305.

McConnaughey, T. & McRoy, C. 1979. Food-web structure and the fractionation of carbon isotopes in the Bering Sea. Mar. Biol. 53(3), 257–262.

McEachran, J. D., Boesch, D. F. & Musick, J. A. 1976. Food division within two sympatric species-pairs of skates (Pisces: Rajidae). Mar. Biol. 35(4), 301–317.

McMillan, D. G., Morse, W. W. & Center, N. F. S. 1999. Spiny dogfish, *Squalus acanthias*, life history and habitat characteristics. Essential fish habitat source document, NOAA Tech. Mem. 150 pp.

Methratta, E. T. & Link, J. S. 2007. Ontogenetic variation in habitat association for four groundfish species in the Gulf of Maine - Georges Bank region. Mar. Ecol. Prog. Ser. 338, 169–181.

Meyer, M. & Smale, M. J. 1991. Predation patterns of demersal teleosts from the Cape south and west coasts of South Africa. 2. benthic and epibenthic predators. S. Afr. J. Mar. Sci. 11(1), 409–442.

Meyer, A. & Zardoya, R. 2003. Recent advances in the (molecular) phylogeny of vertebrates. Annu. Rev. Ecol. Evol. Syst. 34(1), 311–338.

Michener, R. H. & Kaufman, L. 2007. Stable isotope ratios as tracers in marine food webs: an update. In: Stable isotopes in ecology and environmental science. 2nd ed. Eds Michener, R. H., Lajtha, K. John Wiley & Sons, New Jersey. pp. 238–282.

Michener, R. & Schell, D. 1994. Stable isotope ratios as tracers in marine and aquatic food webs. In: Stable isotopes in ecology and environmental science. 1st ed. Eds Lajtha, K., Michener, R. Blackwell Scientific Publications, Oxford. pp. 138 -157.

Miller, T. W., Brodeur, R. D. & Rau, G. H. 2008. Carbon stable isotopes reveal relative contribution of shelf-slope production to the Northern California Current pelagic community. Limnol. Oceanogr. 53(4), 1493–1503.

Minagawa, M. & Wada, E. 1984. Stepwise enrichment of ^{15}N along food chains: further evidence and the relation between ^{15}N and animal age. Geochim. Cosmochim. Acta 48(5), 1135–1140.

Montoya, J., Carpenter, E. & Capone, D. 2002. Nitrogen fixation and nitrogen isotope abundances in zooplankton of the oligotrophic North Atlantic. Limnol. Oceanogr. 47(6), 1617–1628.

Morato, T., Solà, E., Grós, M. P. & Menezes, G. 2003. Diets of thornback ray (*Raja clavata*) and tope shark (*Galeorhinus galeus*) in the bottom longline fishery of the Azores, northeastern Atlantic. Fish. Bull. 101(3), 590–602.

Morato, T., Watson, R., Pitcher, T. & Pauly, D. 2006. Fishing down the deep. Fish Fish. 7(1), 24–34.

Motta, P. J., Tricas, T. C. & Summers, R. 1997. Feeding mechanism and functional morphology of the jaws of the lemon shark *Negaprion brevirostris* (Chondrichthyes, Carcharhinidae). J. Exp. Biol. 200(21), 2765–80.

Mourier, J., Vercelloni, J. & Planes, S. 2012. Evidence of social communities in a spatially structured network of a free-ranging shark species. Anim. Behav. 83(2), 389–401.

Mulas, A., Bellodi, A., Cau, A. L., Gastoni, A., Locci, I. & Follesa, M. C. 2011. Trophic interactions among chondrichthyans in the central-western Mediterranean Sea. Biol. Mar. Mediterr. 18(1), 81–84.

Musick, J. 2011. Chondrichthyan reproduction. In: Reproduction in marine fishes: Evolutionary patterns and innovations.Eds Cole, K. University of California Press, California. pp. 3–19.

Musick, J. & Ellis, J. 2005. Reproductive evolution of chondrichthyans. In: Reproductive biology and phylogeny of Chondrichthyes. Eds Hamlett, W. Science Press, New Hampshire. pp. 45–79.

Navarro, J., Cardador, L., Fernández, Á. M., Bellido, J. M. & Coll, M. 2016. Differences in the relative roles of environment, prey availability and human activity in the spatial distribution of two marine mesopredators living in highly exploited ecosystems. J. Biogeogr. 43(3), 440–450.

Navia, A. F., Mejía-Falla, P. A. & Giraldo, A. 2007. Feeding ecology of elasmobranch fishes in coastal waters of the Colombian Eastern Tropical Pacific. BMC Ecol. 7(1), 8.

Naylor, G., Caira, J., Jensen, K., Rosana, K., White, W. & Last, P. 2012. A DNA sequence–based approach to the identification of shark and ray species and its implications for global elasmobranch diversity and parasitology. Bull. Am. museum Nat. Hist. 1992, 1–262.

Naylor, G., Ryburn, J., Fedrigo, O. & López, J. 2005. Phylogenetic relationships among the major lineages of modern elasmobranchs. In: Reproductive Biology and Phylogeny of Chondrichthyans: Sharks, Skates, Rays and Chimaeras'. 3rd ed. Eds Hamlett, W., Jamiesin, B. Science Publishers, New Hampshire. pp. 1-25.

Nelson, J. S., Grande, T. C. & Wilson, M. V/ H. 2016. Fishes of the World. 5th ed. John Wiley & Sons, New Jersey. 752 pp.

Newsome, S. D., Clementz, M. T. & Koch, P. L. 2010. Using stable isotope biogeochemistry to study marine mammal ecology. Mar. Mamm. Sci. 26(3), 509–572.

Norse, E., Brooke, S., Cheung, W., Clark, M., Ekeland, I., Froese, R., Gjerde, K., Haedrich, R., Heppell, S., Morato, T. & Morgan, L. 2012. Sustainability of deep-sea fisheries. Mar. Policy 36(2), 307–320.

O'Shea, O. R. 2012. The Ecology and Biology of Stingrays (Dasyatidae) at Ningaloo Reef, Western Australia. PhD book, Murdoch University, Australia.

Oddone, M. C., Paesch, L. & Norbis, W. 2010. Size structure, abundance and preliminary information on the reproductive parameters of the shortspine spurdog (*Squalus mitsukurii*) in the Argentinean-Uruguayan Common Fishing Zone from the mid-1990s. J. Northwest Atl. Fish. Sci. 43, 13–26.

Oliver, S., Braccini, M., Newman, S. J. & Harvey, E. S. 2015. Global patterns in the bycatch of sharks and rays. Mar. Policy 54, 86–97.

Pajuelo, J. G., García, S., Lorenzo, J. M. & González, J. A. 2011. Population biology of the shark, *Squalus megalops*, harvested in the central-east Atlantic Ocean. Fish. Res. 108(1), 31–41.

Papastamatiou, Y. P. 2008. Movement patterns, foraging ecology and digestive physiology of blacktip reef sharks, *Carcharhinus melanopterus*, at Palmyra Atoll: a predator dominated ecosystem. PhD book, University of Hawaii, United States.

Papastamatiou, Y. P., Caselle, J. E., Friedlander, A. M. & Lowe, C. G. 2009. Distribution, size frequency, and sex ratios of blacktip reef sharks *Carcharhinus melanopterus* at Palmyra Atoll: A predator-dominated ecosystem. J. Fish Biol. 75(3), 647–654.

Papastamatiou, Y. P., Wetherbee, B. M., Lowe, C. G. & Crow, G. L. 2006. Distribution and diet of four species of carcharhinid shark in the Hawaiian Islands: Evidence for resource partitioning and competitive exclusion. Mar. Ecol. Prog. Ser. 320, 239–251.

Parkins, C. 1993. Stable carbon and nitrogen isotope ratios in the shallow-water Cape hake, *Merluccius capensis* (Castelnau) as indicators of trophic position and diet on the west and south coasts of South Africa. MSc. book, University of Cape Town, South Africa.

Payne, A., Augustyn, C. & Leslie, R. 1985. Biomass index and catch of Cape hake from random stratified sampling cruises in Division 1.6 during 1984. Colln Sci. Pap. int. Commn SE Atl. Fish 12 (2), 99–123.

Petersen, S., Nel, D., Ryan, P. G. & Underhill, L. 2008a. Understanding and mitigating vulnerable and longline fisheries. WWF South Africa Report Series - 2008/Marine/002. WWF South Africa, Cape Town. 225 pp.

Petersen, S., Duncan, J. A., Omardien, A., Betts, M., & Johnson, A. 2015. A decade of implementing an Ecosystem Approach to Fisheries for Southern African fisheries. WWF South Africa Report Series – 20015/Marine/001. WWF South Africa, Cape Town. 42 pp.

Peterson, C. T. 2014. Distribution and abundance, community structure, and trophic ecology of sharks and teleost fishes in the Florida Big Bend. MSc. book, Florida State University, United States.

Peterson, B. & Fry, B. 1987. Stable isotopes in ecosystem studies. Annu. Rev. Ecol. Syst. 18(1), 293–320.

Pethybridge, H. 2010. Ecology and physiology of deepwater chondrichthyans off southeast Australia: mercury,.stable isotope and lipid analysis. PhD book, University of Tasmania, Australia.

Pethybridge, H., Daley, R. K. & Nichols, P. D. 2011. Diet of demersal sharks and chimaeras inferred by fatty acid profiles and stomach content analysis. J. Exp. Mar. Bio. Ecol. 409(1-2), 290–299.

Pianka, E. R. 1981. Competition and niche theory. Theor. Ecol. 8, 167–196.

Pierce, S.J. & Norman, B. 2016. *Rhincodon typus. The IUCN Red List of Threatened Species* 2016: e.T19488A2365291. https://dx.doi.org/10.2305/IUCN.UK.2016-1.RLTS.T19488A2365291.en.

Pikitch, E. K., Chapman, D. D., Babcock, E. a, & Shivji, M. S. 2005. Habitat use and demographic population structure of elasmobranchs at a Caribbean atoll (Glover's Reef,

Belize). Mar. Ecol. Prog. Ser. 302, 187–197.

Pita, R., Mira, A. & Beja, P. 2011. Assessing habitat differentiation between coexisting species: The role of spatial scale. Acta Oecol. 37(2), 124–132.

Platell, M. & Potter, I. 2001. Partitioning of food resources amongst 18 abundant benthic carnivorous fish species in marine waters on the lower west coast of Australia. J. Exp. Mar. Biol. Ecol. 261(1), 31–54.

Platell, M. E., Potter, I. C. & Clarke, K. R. 1998. Resource partitioning by four species of elasmobranchs (Batoidea: Urolophidae) in coastal waters of temperate Australia. Mar. Biol. 131(4), 719–734.

Polunin, N. V. C., Pawsey, W. E., Cartes, J. E., Pinnegar, J. K. & Moranta, J. 2001. Feeding relationships in Mediterranean bathyal assemblages elucidated by stable nitrogen and carbon isotope data. Mar. Ecol. Prog. Ser. 220, 13–23.

Post, D. M. 2002. Using stable isotopes to estimate trophic position: models, methos, and assumptions. Ecology 83(3), 703–718.

Post, D. M., Layman, C. A., Arrington, D. A., Takimoto, G., Quattrochi, J. & Montaña, C. G. 2007. Getting to the fat of the matter: Models, methods and assumptions for dealing with lipids in stable isotope analyses. Oecologia 152(1), 179–189.

Probyn, T., Mitchell-Innes, B., Brown, P., Hutchings, L. & Carter, R. 1994. A review of primary production and related processes on the Agulhas Bank. S. Afr. J. Sci. 90, 166–173.

Pyke, G. 1984. Optimal foraging theory - a critical review. Annu. Rev. Ecol. Syst. 15(1), 523–575.

Quigley, J. 19285. Reactions of an elasmobranch (*Squalus sucklii*) to variations in the salinity of the surrounding medium. Biol. Bull. 54(2), 165–190.

Rau, G., Takahashi, T. & DesMarais, D. 1989. Latitudinal variations in plankton $\delta^{13}C$: implications for CO_2 and productivity in past oceans. Nature 341(6242), 516–518.

Reed, J. R., Kerwath, S. E. & Attwood, C. G. 2017. Analysis of bycatch in the South African midwater trawl fishery for horse mackerel *Trachurus capensis* based on observer data. Afr. J. Mar. Sci. 39(3), 279–291.

Regan, C. 1908. On the sharks of the family Squaloidae. VII—A synopsis of the sharks of the family Squaloidae. Ann. Mag. Nat. Hist. 2, 39–57.

Reid, W. D. K., Sweeting, C. J., Wigham, B. D., McGill, R. A. R. & Polunin, N. V. C. 2013. High variability in spatial and temporal size-based trophodynamics of deep-sea fishes from the Mid-Atlantic Ridge elucidated by stable isotopes. Deep. Res. Part II Top. Stud. Oceanogr. 98, 412–420.

Reum, J. C. P. & Essington, T. E. 2013. Spatial and seasonal variation in $\delta^{15}N$ and $\delta^{13}C$ values in a mesopredator shark, *Squalus suckleyi*, revealed through multitissue analyses. Mar. Biol. 160(2), 399–411.

Richardson, A., Maharaj, G., Compagno, L., Leslie, R., Ebert, D. & Gibbons, M. 2000. Abundance, distribution, morphometrics, reproduction and diet of the Izak catshark. J. Fish Biol. 56(3), 552–576.

Rinewalt, C. S. 2007. Diet and ecomorphology of the sandpaper skate, *Bathyraja kincaidii* (Garman, 1908) from the eastern North Pacific. MSc. book, California State University, United States.

Ritchie, E. G. & Johnson, C. N. 2009. Predator interactions, mesopredator release and biodiversity conservation. Ecol. Lett. 12(9), 982–998.

Ross, S. T. 1986. Resource partitioning in fish assemblages: a review of field studies. Copeia 1986, 352–388.

Ruocco, N. L., Lucifora, L. O., Diaz de Astarloa, J. M., Menni, R. C., Mabragana, E. & Giberto, D. A. 2012. From coexistence to competitive exclusion: can overfishing change the outcome of competition in skates (Chondrichthyes, Rajidae)? Lat. Am. J. Aquat. Res. 40(1), 102–112.

Sale, P. F. 1974. Overlap in resource use, and interspecific competition. Oecologia 17(3), 245-256.

Schiffman, D., Gallagher, A., Boyle, M., Hammerschlag-Peyer, C. & Hammerschlag, N. 2012. Stable isotope analysis as a tool for elasmobranch conservation: a primer for non-specialists. Mar. Freshw. Res. 63(7), 635–643.

Schmidt, S. N., Olden, J. D., Solomon, C. T. & Vander Zanden, M. J. 2007. Quantitative approaches to the analysis of stable isotope food web data. Ecology 88(11), 2793–2802.

Schoener, T. W. 1974. Resource partitioning in ecological communities. Science 185 (4145),

185(4145), 27–39.

Schoener, T., Ross, S., Myers, R., Baum, J., Shepard, T., Powers, S., Peterson, C., Keina, M., Norse, E., Brooke, S., Cheung, W., Clark, M., Ekeland, I., Froese, R., Gjerde, K., Haedrich, R., Heppell, S., Morato, T., Morgan, L., Pauly, D., Sumaila, R., Watson, R., Vaudo, J., Heithaus, M., Dicken, M., Hussey, N., Christiansen, H., Smale, M., Nkabi, N., Cliff, G., Wintner, S., Caut, S., Jowers, M., Michel, L., Lepoint, G., Fisk, A., Costa, T., Thayer, J., Mendes, L., Bizzarro, J., Robinson, H., Rinewalt, C. & Ebert, D. 2012a. Sustainability of deep-sea fisheries. ICES J. Mar. Sci. 53, 1846–1850.

Sherwood, G. D. & Rose, G. A. 2005. Stable isotope analysis of some representative fish and invertebrates of the Newfoundland and Labrador continental shelf food web. Estuar. Coast. Shelf Sci. 63(4), 537–549.

Shiffman, D. S., Frazier, B. S., Kucklick, J. R., Abel, D., Brandes, J. & Sancho, G. 2014. Feeding ecology of the sandbar shark in south Carolina estuaries revealed through δ^{13}C and δ^{15}N stable isotope analysis. Mar. Coast. Fish. 6(1), 156–169.

Shipley, O. N., Brooks, E. J., Madigan, D. J., Sweeting, C. J. & Dean Grubbs, R. 2017. Stable isotope analysis in deep-sea chondrichthyans: recent challenges, ecological insights, and future directions. Rev. Fish Biol. Fish. 27(3), 481–497.

Simpfendorfer, C. A., Freitas, G. G., Wiley, T. R. & Heupel, M. R. 2005. Distribution and habitat partitioning of immature bull sharks (*Carcharhinus leucas*) in a Southwest Florida estuary. Estuaries(1) 28, 78–85.

Simpfendorfer, C. A., Goodreid, A. B. & Mcauley, R. B. 2001. Size, sex and geographic variation in the diet of the tiger shark, *Galeocerdo cuvier*, from Western Australian waters. Mar. Freshw. Res. 61, 37–46.

Simpfendorfer, C. & Kyne, P. 2009. Limited potential to recover from overfishing raises concerns for deep-sea sharks, rays and chimaeras. Environ. Conserv. 36, 97–103.

Sims, D. W. 2003. Tractable models for testing theories about natural strategies: foraging behaviours and habitat selection of free ranging sharks. J. Fish Biol. 63, 53–73.

Sims, D., Nash, J. & Morritt, D. 2001. Movements and activity of male and female dogfish in a tidal sea lough: alternative behavioural strategies and apparent sexual segregation. Mar. Biol. 139(6), 1165–1175.

Smale, M. J. & Compagnon, L. J. V. 1997. Life history and diet of two southern African smoothhound sharks, *Mustelus mustelus* (Linnaeus, 1758) and *Mustelus palumbes* Smith, 1957 (Pisces: Triakidae). S. Afr. J. Mar. Sci. 18(1), 229–248.

Speers-Roesch, B., Brauner, C. J., Farrell, A. P., Hickey, A. J. R., Renshaw, G. M. C., Wang, Y. S. & Richards, J. G. 2012. Hypoxia tolerance in elasmobranchs. II. Cardiovascular function and tissue metabolic responses during progressive and relative hypoxia exposures. J. Exp. Biol. 215(1), 103–114.

Springer, S. 1967. Social organization of shark populations. In: Sharks, skates and rays. Eds Gilbert, P., Mathewson, R., Rall, D. John Hopkins Press, Baltimore. pp. 149 - 174.

Stevens, J., Bonfil, R., Dulvy, N. & Walker, P. 2000. The effects of fishing on sharks, rays, and chimaeras (chondrichthyans), and the implications for marine ecosystems. ICES J. Mar. Sci. 57(3), 476–494.

Takai, N., Mishima, Y., Yorozu, A. & Hoshika, A. 2002. Carbon sources for demersal fish in the western Seto Inland Sea, Japan, examined by $\delta^{13}C$ and $\delta^{15}N$ analyses. Limnol. Oceanogr. 47(3), 730–741.

Taylor, S., Sumpton, W. & Ham, T. 2011. Fine-scale spatial and sea sonal partitioning among large sharks and other elasmobranchs in south-eastern Queensland, Australia. Mar. Freshw. Res. 62(6), 638–647.

Thomas, C. J. & Cahoon, L. B. 1993. Stable isotope analyses differentiate between different trophic pathwasy supporting rocky-reef fishes. Mar. Ecol. Prog. Ser. 95, 19–24.

Thompson, J. & Springer, S. 1965. Sharks, Skates, Ray, and Chimeras. U.S. Department of the Interior, Fish and Wildlife Service, Bureau of Commercial Fisheries, California. 18 pp.

Tillett, B. J., Meekan, M. G. & Field, I. C. 2014. Dietary overlap and partitioning among three sympatric carcharhinid sharks. Endanger Species Res. 25(3), 283–293.

Tilley, A. 2011. Functional ecology of the southern stingray, *Dasyatis americana*. PhD book, Bangor University, Wales.

Tilley, A., Lopez-Angarita, J. & Turner, J. R. 2013. Diet reconstruction and resource partitioning of a Caribbean marine mesopredator using stable isotope Bayesian modelling. PLoS One 8(11), e79560.

Tirasin, E. M. & Jørgensen, T. 1999. An evaluation of the precision of diet description. Mar. Ecol. Prog. Ser. 182, 243–252.

Trueman, C. N., Johnston, G., O'Hea, B. & MacKenzie, K. M. 2014. Trophic interactions of fish communities at midwater depths enhance long-term carbon storage and benthic production on continental slopes. Proc. R. Soc. B Biol. Sci. 281(1787), 20140669.

Valesini, F. J., Potter, I. C., Platell, M. E. & Hyndes, G. A. 1997. Ichthyofaunas of a temperate estuary and adjacent marine embayment. Implications regarding choice of nursery area and influence of environmental changes. Mar. Biol. 128(2), 317–328.

Valls, M., Quetglas, A., Ordines, F. & Moranta, J. 2011. Feeding ecology of demersal elasmobranchs from the shelf and slope off the Balearic Sea (western Mediterranean). Sci. Mar. 75(4), 633–639.

Valls, M., Rueda, L. & Quetglas, A. 2017. Feeding strategies and resource partitioning among elasmobranchs and cephalopods in Mediterranean deep-sea ecosystems. Deep Sea Res. Part I Oceanogr. Res. Pap. 128, 28–41.

van der Elst, R. 1993. A guide to the common sea fishes of southern Africa. Struik Publishers, Cape Town. 398 pp.

van der Heever, G. M. 2017. Intra- and interspecific variability in the distribution patterns and diet of the two most common catsharks caught in demersal trawls off the West and South coasts of South Africa: Evidence for habitat and resource partitioning? MSc. thesis, University of the Western Cape, South Africa.

van der Lingen, C. D. & Miller, T. W. 2014. Spatial, ontogenetic and interspecific variability in stable isotope ratios of nitrogen and carbon of *Merluccius capensis* and *Merluccius paradoxus* off South Africa. J. Fish Biol. 85(2), 456–472.

Varela, J. L., Rodríguez-Marín, E. & Medina, A. 2013. Estimating diets of pre-spawning Atlantic bluefin tuna from stomach content and stable isotope analyses. J. Sea Res. 76, 187–192.

Vaudo, J. J. & Heithaus, M. R. 2011. Dietary niche overlap in a nearshore elasmobranch mesopredator community. Mar. Ecol. Prog. Ser. 425, 247–260.

Veríssimo, A., McDowell, J. R. & Graves, J. E. 2011. Population structure of a deep-water squaloid shark, the Portuguese dogfish (*Centroscymnus coelolepis*). ICES J. Mar. Sci.

68(3), 555–563.

Veríssimo, A., Zaera-Perez, D., Leslie, R., Iglésias, S. ., Séret, B., Grigoriou, P., Sterioti, A., Gubili, C., Barría, C., Duffy, C. & Hernández, S. 2017. Molecular diversity and distribution of eastern Atlantic and Mediterranean dogfishes *Squalus* highlight taxonomic issues in the genus. Zool. Scripta 46(4), 414–428.

Viana, S. & de Carvalho, M. 2016. Redescription of *Squalus acutipinnis* Regan, 1908, a valid species of spiny dogfish from Southern Africa (Chondrichthyes: Squaliformes: Squaloidae). Copeia 104(2), 539–553.

Viana, S. T. d. F. L., de Carvalho, M. R. & Ebert, D. A. 2017. *Squalus bassi* sp. nov., a new long-snouted spurdog (Chondrichthyes: Squaliformes: Squaloidae) from the Agulhas Bank. J. Fish Biol. 91(4), 1178–1207.

Vögler, R., Milessi, A. C. & Duarte, L. O. 2009. Changes in trophic level of *Squatina guggenheim* with increasing body length: relationships with type, size and trophic level of its prey. Environ. Biol. Fishes 84(1), 41–52.

Waite, A., Muhling, B., Holl, C., Beckley, L., Montoya, J., Strzelecki, J., Thompson, P. & Pesant, S. 2007. Food web structure in two counter-rotating eddies based on $\delta^{15}N$ and $\delta^{13}C$ isotopic analyses. Deep Sea Res. Part I Top. Stud. Oceanogr. 54(8-10), 1055–1075.

Walker, T. 2020. Reproduction of Chondrichthyans. In: Reproduction in Aquatic Animals: From Basic Biology to Aquaculture Technology.Eds Yoshida, M., Asturiano, J. Springer International Publishing, Singapore. pp. 193–223.

Waser, N., Harrison, W., Head, E., Nielsen, B., Lutz, V. & Calvert, S. 2000. Geographic variations in the nitrogen isotope composition of surface particulate nitrogen and new production across the North Atlantic Ocean. Deep Sea Res. Part I Oceanogr. Res. Pap. 47(7), 1207–1226.

Watson, G. & Smale, M. J. 1998. Reproductive biology of shortnose spiny dogfish, *Squalus megalops*, from the Agulhas Bank, South Africa. Mar. Freshw. Res. 49(7), 695–703.

Watson, G. & Smale, M. J. 1999. Age and growth of the shortnose spiny dogfish *Squalus megalops* from the Agulhas Bank, South Africa. S. Afr. J. Mar. Sci. 21, 9–18.

Wetherbee, B. & Cortés, E. 2004. Food consumption and feeding habits. In: Biology of sharks and their relatives.1ˢᵗ ed. Eds Carrier, J., Musick, J., Heithaus, M. CRC Press, Florida. pp. 225–246.

Wetherbee, B. M., Cortés, E. & Bizzarro, J. J. 2012. Food Consumption and Feeding Habits. In: Biology of sharks and their relatives. 2ⁿᵈ ed. Eds Carrier, J., Musick, J. A., Heithaus, M. R. crc Press, Florida. pp. 239–264.

Wetherbee, B. M., Gruber, H. S. & Cortés, E. 1990. Diet, feeding habitat, digestion, and consumption in sharks, with special reference to the lemon shark, *Negaprion brevirostris*. Elasmobranchs as living resources. NOAA Tech Rep, NM FS, 90(1), 29–47.

White, W. T., Ebert, D. A., Naylor, G. J. P., Ho, H. C., Clerkin, P., Veríssimo, A. & Cotton, C. F. 2013. Revision of the genus *Centrophorus* (squaliformes: Centrophoridae): Part 1-redescription of *Centrophorus granulosus* (bloch & schneider), a senior synonym of *Cacus garman* and *C. niaukang teng*. Zootaxa 3752(1), 35–72.

White, W. & Iglésias, S. 2011. *Squalus formosus*, a new species of spurdog shark (Squaliformes: Squaloidae), from the western North Pacific Ocean. J. Fish Biol. 79(4), 954–968.

White, W., Platell, M. & Potter, I. 2004. Comparisons between the diets of four abundant species of elasmobranchs in a subtropical embayment: implications for resource partitioning. Mar. Biol. 144(3), 439–448.

White, W. T. & Potter, I. C. 2004. Habitat partitioning among four elasmobranch species in nearshore, shallow waters of a subtropical embayment in Western Australia. Mar. Biol. 145(5), 1023–1032.

Wiley, T. R. & Simpfendorfer, C. A. 2007. The ecology of elasmobranchs occuring in the Everglades National Park, Florida: Implications for conservation and management. Bull. Mar. Sci. 80(1), 171–189.

Wilson, C. & Seki, M. 1994. Biology and population characteristics of *Squalus mitsukurii* from a seamount in the central North Pacific Ocean. Fish. Bull. 92(4), 851–564.

Woodland, R. J., Secor, D. H. & Wedge, M. E. 2011. Trophic resource overlap between small elasmobranchs and sympatric teleosts in Mid-Atlantic Bight nearshore habitats. Estuaries Coast. 34(2), 391–404.

Worm, B., Davis, B., Kettemer, L., Ward-Paige, C. A., Chapman, D., Heithaus, M. R., Kessel, S. T. & Gruber, S. H. 2013. Global catches, exploitation rates, and rebuilding options for sharks. Mar. Policy 40, 194–204.

Wyatt, A., Waite, A. & Humphries, S. 2012. Stable isotope analysis reveals community-level variation in fish trophodynamics across a fringing coral reef. Coral Reefs 31(4), 1029–1044.

Yamaguchi, A. & Taniuchi, T. 2000. Food variations and ontogenetic dietary shift of the starspotted-dogfish *Mustelus manazo* at five locations in Japan and Taiwan. Fish. Sci. 66(6), 1039–1048.

Yick, J. L., Tracey, S. R. & White, R. W. G. 2011. Niche overlap and trophic resource partitioning of two sympatric batoids co-inhabiting an estuarine system in southeast Australia. J. Appl. Ichthyol. 27(5), 1272–1277.

Vander Zanden, M. & Vadeboncoeur, Y. 2002. Fishes as integrators of benthic and pelagic food webs in lakes. Ecology 83(8), 2152–2161.

Zar, J. H. E. 1999. Biostatistical analysis. Prentice Hall, New Jersey. 663 pp.

Zar, J. H. E. 2010. Biostatistical Analysis. 5th ed. Pearson Prentice Hall International Editions, New Jersey. 994 pp.

Zimmer, A. M. & Wood, C. M. 2014. Exposure to acute severe hypoxia leads to increased urea loss and disruptions in acid-base and ionoregulatory balance in dogfish sharks (*Squalus acanthias*). Physiol. Biochem. Zool. 87(5), 623–639.

Appendices

Appendix 1: The full stomach content dataset for small *Squalus acutipinnis* sharks sampled at the 250 m depth bin along the West coast of South Africa. A total of 7 stomachs were analysed

Taxon	%N	%W	% FO	% IRI
Phylum Mollusca	**42.9**	**10.3**	**42.9**	**39.5**
Unidentified cephalopod	42.9	10.3	42.9	39.5
Phylum Chordata: Osteichthyes	**57.1**	**89.7**	**57.1**	**60.5**
Pegusa nasuta	14.3	84.1	14.3	24.4
Unidentified teleost	42.9	5.6	42.9	36.1
SUM	**100.0**	**100.0**	**100.0**	**100.0**

Appendix 2: The full stomach content dataset for small *Squalus acutipinnis* sharks sampled at the 50 m depth bin along the South coast of South Africa. A total of 26 stomachs were sampled

Species	%N	%W	% FO	% IRI
Phylum Annelida	**76.2**	**58.5**	**69.2**	**92.6**
Unidentified polychaete	76.2	58.5	69.2	92.6
Phylum Arthropoda	**2.4**	**0.5**	**3.8**	**0.1**
Unidentified crustacean	2.4	0.5	3.8	0.1
Phylum Echinodermata	**2.4**	**0.8**	**3.8**	**0.1**
Ophiothrix sp.	2.4	0.8	3.8	0.1
Phylum Chordata: Osteichthyes	**19.0**	**40.2**	**23.1**	**7.1**
Engraulis encrasicolus	14.3	19.8	15.4	5.2
Trachurus capensis	4.8	20.4	7.7	1.9
SUM	**100.0**	**100.0**	**100.0**	**100.0**

Appendix 3: The full stomach content dataset for small *Squalus acutipinnis* sharks sampled at the 150 m depth bin along the South coast of South Africa. A total of 35 stomachs were analysed

Species	%N	%W	% FO	% IRI
Phylum Annelida	**0.4**	**0.4**	**2.9**	**0.1**
Unidentified polychaete	0.4	0.4	2.9	0.1
Phylum Arthropoda	**90.0**	**35.4**	**40.0**	**56.3**
Caridea	0.4	0.2	2.9	0.1
Funchalia woodwardi	0.4	2.5	2.9	0.2
Pasiphaea spp. *1*	71.1	27.4	17.1	51.3
Unidentified amphipod	16.8	3.5	5.7	3.5
Unidentified crustacean	1.4	1.8	11.4	1.1
Phylum Mollusca	**0.7**	**1.9**	**5.7**	**0.4**
Unidentified cephalopod	0.7	1.9	5.7	0.4
Phylum Chordata: Osteichthyes	**8.2**	**49.8**	**45.7**	**40.9**
Engraulis encrasicolus	3.2	18.9	11.4	7.7
Paracallionymus costatus	0.4	0.9	2.9	0.1
Unidentified teleost	4.6	30.0	31.4	33.1
Unidentified semi-digested material	**0.7**	**12.6**	**5.7**	**2.3**
Unidentified specimen	0.7	12.6	5.7	2.3
SUM	**100.0**	**100.0**	**100.0**	**100.0**

Appendix 4: The full stomach content dataset for medium *Squalus acutipinnis* sharks sampled at the 50 m depth bin along the West coast of South Africa. A total of 10 stomachs were analysed

Species	%N	%W	% FO	% IRI
Phylum Annelida	**45.5**	**16.7**	**70.0**	**55.1**
Unidentified polychaete	45.5	16.7	70.0	55.1
Phylum Arthropoda	**18.2**	**2.1**	**40.0**	**3.9**
Caridea	9.1	1.2	20.0	2.6
Goneplax angulata	4.5	0.8	10.0	0.7
Mursia cristiata	4.5	0.0	10.0	0.6
Phylum Mollusca	**9.1**	**19.6**	**20.0**	**7.3**
Unidentified cephalopod	9.1	19.6	20.0	7.3
Phylum Chordata: Osteichthyes	**27.3**	**61.6**	**60.0**	**33.8**
Engraulis encrasicolus	13.6	52.0	30.0	24.9
Unidentified teleost	13.6	9.6	30.0	8.8
SUM	100.0	100.0	190.0	100.0

Appendix 5: The full stomach content dataset for medium *Squalus acutipinnis* sharks sampled at the 150 m depth bin along the West coast of South Africa. A total of 21 stomachs were analysed

Species	%N	%W	% FO	%IRI
Phylum Arthropoda	**32.1**	**2.5**	**33.3**	**16.9**
Mysida	21.4	0.8	19.0	13.7
Malacostraca	7.1	0.9	9.5	2.5
Pterygosquilla armata capensis	3.6	0.9	4.8	0.7
Phylum Mollusca	**17.9**	**3.8**	**23.8**	**11.5**
Unidentified cephalopod	14.3	3.4	19.0	10.9
Unidentified mollusc	3.6	0.4	4.8	0.6
Phylum Chordata: Osteichthyes	**46.4**	**93.2**	**57.1**	**71.0**
Callionymidae	7.1	8.5	4.8	2.4
Engraulis encrasicolus	7.1	11.2	9.5	5.6
Etrumeus whiteheadi	7.1	59.2	9.5	20.5
Unidentified teleost	25.0	14.4	33.3	42.5
Unidentified semi-digested material	**3.6**	**0.4**	**4.8**	**0.6**
Unidentified invertebrate	3.6	0.4	4.8	0.6
SUM	**100.0**	**100.0**	**119.0**	**100.0**

Appendix 6: The full stomach content dataset for medium *Squalus acutipinnis* sharks sampled at the 250 m depth bin along the West coast of South Africa. A total of 12 stomachs were analysed

Species	%N	%W	% FO	% IRI
Phylum Arthropoda	**33.3**	**1.3**	**16.7**	**12.8**
Unidentified crustacean	33.3	1.3	16.7	12.8
Phylum Mollusca	**14.3**	**45.0**	**25.0**	**13.4**
Todaropsis eblanae	4.8	41.6	8.3	8.6
Unidentified cephalopod	9.5	3.5	16.7	4.8
Phylum Chordata: Osteichthyes	**42.9**	**51.5**	**66.7**	**69.5**
Engraulis encrasicolus	9.5	38.3	16.7	17.7
Unidentified teleost	33.3	13.3	50.0	51.8
Unidentified semi-digested material	**9.5**	**2.1**	**16.7**	**4.3**
Unidentified invertebrate	9.5	2.1	16.7	4.3
SUM	100.0	100.0	125.0	100.0

Appendix 7: The full stomach content dataset for medium *Squalus acutipinnis* sharks sampled at the 50 m depth bin along the South coast of South Africa. A total of 84 stomachs were analysed

Species	%N	%W	% FO	%IRI
Phylum Annelida	**19.1**	**6.2**	**25.0**	**22.5**
Unidentified polychaete	19.1	6.2	25.0	22.5
Phylum Arthropoda	**17.8**	**4.7**	**28.6**	**2.8**
Decapoda	1.9	1.1	3.6	0.4
Goneplax angulata	3.8	1.3	6.0	1.1
Malacostraca	0.6	0.4	1.2	0.0
Mursia cristiata	0.6	0.3	1.2	0.0
Mysida	1.3	0.0	1.2	0.1
Nassarius vinctus	1.9	0.6	2.4	0.2
Parapaguridae	1.3	0.4	2.4	0.1
Penaeidae	1.3	0.3	2.4	0.1
Unidentified amphipod	1.9	0.1	2.4	0.2
Unidentified crustacean	2.5	0.2	4.8	0.5
Unidentified isopod	0.6	0.0	1.2	0.0
Phylum Mollusca	**6.4**	**25.5**	**11.9**	**3.8**
Loligo reynaudii	1.9	19.0	3.6	2.7
Unidentified cephalopod	3.2	0.9	6.0	0.9
Unidentified mollusc	0.6	0.1	1.2	0.0
Unidentified octopus	0.6	5.5	1.2	0.3
Phylum Chordata: Osteichthyes	**54.8**	**62.7**	**77.4**	**70.6**
Austroglossus pectoralis	1.9	1.9	2.4	0.3
Engraulis encrasicolus	14.0	16.6	20.2	22.0
Etrumeus whiteheadi	4.5	8.9	4.8	1.0
Genypterus capensis	0.6	0.8	1.2	0.1
Merluccius capensis	1.3	0.5	2.4	0.2
Merluccius sp.	1.3	0.3	1.2	0.1
Paracallionymus costatus	1.9	1.0	3.6	0.4
Sardinops sagax	1.9	5.8	3.6	1.0
Scomberesox saurus scomberoides	0.6	0.3	1.2	0.0
Trachurus capensis	5.7	8.9	6.0	3.1
Unidentified teleost	21.0	17.7	31.0	42.5
Unidentified semi-digested material	**1.9**	**0.8**	**3.6**	**0.3**
Unidentified specimen	1.9	0.8	3.6	0.3
SUM	**100.0**	**100.0**	**146.4**	**100.0**

Appendix 8: The full stomach content dataset for medium *Squalus acutipinnis* sharks sampled at the 150 m depth bin along the South coast of South Africa. A total of 159 stomachs were analysed

Species	% N	% W	% FO	% IRI
Phylum Annelida	**23.5**	**14.4**	**35.2**	**42.5**
Nereididae	0.2	0.8	0.6	0.0
Unidentified polychaete	23.2	13.6	34.6	42.5
Phylum Sipuncula	**0.2**	**0.4**	**0.6**	**0.0**
Sipuncula	0.2	0.4	0.6	0.0
Phylum Arthropoda	**40.6**	**14.6**	**51.6**	**8.8**
Caridea	2.1	0.4	5.7	0.5
Decapoda	2.1	0.6	5.0	0.5
Euphausiacea	0.7	0.0	1.9	0.0
Funchalia woodwardi	1.2	0.1	1.9	0.1
Malacostraca	0.5	0.1	1.3	0.0
Mursia cristiata	1.4	1.1	2.5	0.2
Mysida	0.7	0.2	1.9	0.1
Parapaguridae	0.9	0.5	2.5	0.1
Pasiphaea spp. 1	6.8	0.2	0.6	0.1
Plesionika martia	0.7	0.5	1.9	0.1
Pterygosquilla armata capensis	2.3	1.1	4.4	0.5
Unidentified amphipod	13.1	0.8	6.9	3.2
Unidentified crustacean	5.6	1.4	10.1	2.4
Unidentified isopod	0.2	0.0	0.6	0.0
Upogebia africana	0.2	0.0	0.6	0.0
Upogebia capensis	0.2	0.0	0.6	0.0
Upogebia sp.	1.6	7.4	3.1	1.0
Phylum Mollusca	**5.9**	**11.6**	**15.7**	**2.5**
Loligo reynaudii	0.7	4.7	1.9	0.3
Ommastrephidae	0.2	3.2	0.6	0.1
Sepia australis	0.2	0.3	0.6	0.0
Teuthida	0.2	0.8	0.6	0.0
Unidentified cephalopod	4.0	1.8	10.7	2.1
Unidentified gastropod	0.2	0.0	0.6	0.0
Unidentified mollusc	0.2	0.8	0.6	0.0
Phylum Hemichordata	**0.2**	**0.2**	**0.6**	**0.0**
Enteropneusta	0.2	0.2	0.6	0.0
Phylum Chordata: Chondrichthyes	**0.2**	**0.6**	**0.6**	**0.0**
Unidentified elasmobranch	0.2	0.6	0.6	0.0
Phylum Chordata: Osteichthyes	**24.9**	**51.7**	**58.5**	**42.2**
Bregmaceros sp.	0.5	0.0	1.3	0.0
Engraulis encrasicolus	3.3	5.3	7.5	2.2
Etrumeus whiteheadi	0.9	6.5	1.3	0.3
Merluccius capensis	0.2	1.2	0.6	0.0
Merluccius sp.	0.9	1.3	2.5	0.2
Paracallionymus costatus	3.8	4.8	8.2	2.3
Sardinops sagax	0.9	10.9	2.5	1.0
Trachurus capensis	0.9	2.8	1.3	0.2
Unidentified teleost	13.4	19.0	33.3	36.0
Unidentified semi-digested material	**4.5**	**6.7**	**10.7**	**4.0**
Unidentified specimen	4.5	6.7	10.7	4.0
SUM	**100.0**	**100.0**	**173.6**	**100.0**

Appendix 9: The full stomach content dataset for large *Squalus acutipinnis* sharks sampled at the 50 m depth bin along the West coast of South Africa. A total of 12 stomachs were analysed

Species	%N	%W	% FO	% IRI
Phylum Annelida	**33.3**	**16.3**	**66.7**	**62.7**
Unidentified polychaete	33.3	16.3	66.7	62.7
Phylum Arthropoda	**25.0**	**21.1**	**50.0**	**8.7**
Caridea	8.3	0.9	16.7	2.9
Goneplax angulata	4.2	0.6	8.3	0.8
Mursia cristiata	4.2	7.5	8.3	1.8
Penaeidae	4.2	3.0	8.3	1.1
Upogebia capensis	4.2	9.0	8.3	2.1
Phylum Mollusca	**4.2**	**1.5**	**8.3**	**0.9**
Unidentified cephalopod	4.2	1.5	8.3	0.9
Phylum Chordata: Osteichthyes	**37.5**	**61.1**	**66.7**	**27.7**
Engraulis encrasicolus	12.5	19.0	25.0	14.9
Merluccius sp.	4.2	7.5	8.3	1.8
Paracallionymus costatus	4.2	1.8	8.3	0.9
Sardinops sagax	8.3	27.1	8.3	5.6
Unidentified teleost	8.3	5.6	16.7	4.4
SUM	**100.0**	**100.0**	**191.7**	**100.0**

Appendix 10: The full stomach content dataset for large *Squalus acutipinnis* sharks sampled at the 150 m depth bin along the West coast of South Africa. A total of 35 stomachs were analysed

Species	%N	%W	% FO	% IRI
Phylum Annelida	**2.8**	**0.6**	**5.7**	**0.6**
Unidentified polychaete	2.8	0.6	5.7	0.6
Phylum Arthropoda	**18.1**	**5.9**	**34.3**	**7.7**
Funchalia woodwardi	6.9	1.2	14.3	3.4
Malacostraca	1.4	0.0	2.9	0.1
Mysida	6.9	3.1	11.4	3.4
Pterygosquilla armata capensis	2.8	1.5	5.7	0.7
Phylum Mollusca	**9.7**	**7.9**	**20.0**	**3.6**
Sepia australis	2.8	3.7	5.7	1.1
Unidentified cephalopod	4.2	3.3	8.6	1.9
Unidentified mollusc	2.8	1.0	5.7	0.6
Phylum Echinodermata	**1.4**	**5.5**	**2.9**	**0.6**
Holanthias sp.	1.4	5.5	2.9	0.6
Phylum Chordata: Osteichthyes	**65.3**	**79.8**	**100.0**	**87.0**
Callionymidae	18.1	13.5	17.1	16.0
Engraulis encrasicolus	9.7	17.8	17.1	13.9
Gnathophis sp.	1.4	5.0	2.9	0.5
Lophiiformes	1.4	0.6	2.9	0.2
Merluccius capensis	1.4	11.0	2.9	1.0
Merluccius sp.	1.4	7.5	2.9	0.7
Sardinops sagax	2.8	2.0	2.9	0.4
Symbolophorus boops	4.2	3.3	5.7	1.3
Trichiurus lepturus	1.4	1.1	2.9	0.2
Unidentified teleost	23.6	18.0	42.9	52.7
Unidentified semi-digested material	**2.8**	**0.3**	**5.7**	**0.5**
Unidentified invertebrate	2.8	0.3	5.7	0.5
SUM		**100.0**	**100.0**	**168.6**	**100.0**

Appendix 11: The full stomach content dataset for large *Squalus acutipinnis* sharks sampled at the 250 m depth bin along the West coast of South Africa. A total of 20 stomachs were analysed

Species	%N	%W	% FO	% IRI
Phylum: Annelida	**2.27**	**0.81**	**4.76**	**0.51**
Unidentified polychaete	2.27	0.81	4.76	0.51
Phylum: Arthropoda	**22.73**	**5.93**	**42.86**	**17.79**
Funchalia woodwardi	9.09	1.37	19.05	6.91
Malacostraca	2.10	0.03	2.91	0.29
Mysida	10.0	4.51	18.05	10.45
Pterygosquilla armata capensis	1.36	0.01	1.85	0.09
Unidentified crustacean	1.17	0.01	1.00	0.05
Phylum Mollusca	**4.55**	**2.20**	**9.52**	**1.11**
Todaropsis eblanae	2.27	2.00	4.76	0.71
Unidentified cephalopod	2.27	0.20	4.76	0.41
Phylum Echinodermata	**2.27**	**8.07**	**4.76**	**1.71**
Holanthias sp.	2.27	8.07	4.76	1.71
Phylum Chordata: Osteichthyes	**65.91**	**82.79**	**100.00**	**78.47**
Callionymidae	20.45	12.51	19.05	21.78
Engraulis encrasiocolis	11.36	17.15	19.05	18.84
Gnathophis sp.	2.27	7.26	4.76	1.58
Merluccius capensis	2.27	16.09	4.76	3.03
Merluccius sp.	2.27	10.89	4.76	2.18
Sardinops sagax	4.55	2.95	4.76	1.24
Lampricoides	6.82	1.21	9.52	2.65
Trichiurus lepturus	2.27	1.61	4.76	0.64
Unidentified teleost	13.64	13.12	28.57	26.53
Unidentified semi-digested material	**2.27**	**0.20**	**4.76**	**0.41**
Unidentified invertebrate	2.27	0.20	4.76	0.41
SUM	**100.00**	**100.00**	**166.67**	**100.00**

Species	%N	%W	% FO	% IRI

Appendix 12: The full stomach content dataset for large *Squalus acutipinnis* sharks sampled at the 350 m depth bin along the West coast of South Africa. A total of 6 stomachs were analysed

Species	%N	%W	% FO	% IRI
Phylum Annelida	**14.3**	**0.2**	**16.7**	**4.3**
Unidentified polychaete	14.3	0.2	16.7	4.3
Phylum Mollusca	**42.9**	**7.9**	**50.0**	**23.3**
Todaropsis eblanae	14.3	6.0	16.7	5.1
Unidentified cephalopod	28.6	1.8	33.3	18.2
Phylum Chordata: Osteichthyes	**42.9**	**92.0**	**50.0**	**72.4**
Clupeidae	14.3	9.5	16.7	5.9
Unidentified teleost	28.6	82.5	33.3	66.5
SUM	**100.0**	**100.0**	**116.7**	**100.0**

Appendix 13: The full stomach content dataset for large *Squalus acutipinnis* sharks sampled at the 50 m depth bin along the South coast of South Africa. A total of 214 stomachs were analysed

Species	% N	% W	% FO	% IRI
Phylum Annelida	**24.5**	**7.4**	**26.6**	**31.5**
Unidentified polychaete	24.5	7.4	26.6	31.5
Phylum Arthropoda	**12.1**	**2.6**	**24.3**	**1.3**
Brachyura	0.2	0.0	0.5	0.0
Caridea	1.0	0.0	2.3	0.1
Decapoda	1.2	0.2	2.8	0.1
Funchalia woodwardi	0.2	0.0	0.5	0.0
Goneplax angulata	2.3	0.4	3.7	0.4
Mursia cristiata	0.2	0.0	0.5	0.0
Mysida	1.2	0.0	1.9	0.1
Parapaguridae	0.8	0.1	1.9	0.1
Penaeidae	0.2	0.0	0.5	0.0
Plesionika martia	0.4	0.0	0.9	0.0
Pterygosquilla armata capensis	1.4	0.4	3.3	0.2
Unidentified amphipod	1.4	0.1	2.8	0.1
Unidentified crustacean	0.6	0.0	1.4	0.0
Upogebia sp.	1.2	1.3	1.4	0.1
Phylum Mollusca	**11.3**	**21.7**	**24.3**	**5.7**
Aplysiidae	0.6	0.1	1.4	0.0
Chiroteuthis mega	0.2	0.0	0.5	0.0
Loligo reynaudii	3.7	9.1	7.9	3.8
Afrololigo mercatoris	0.4	0.1	0.9	0.0
Octopus vulgaris	0.4	1.7	0.9	0.1
Sepia australis	1.6	4.4	2.8	0.6
Sepia sp.	0.2	0.1	0.5	0.0
Sepiolidae	0.2	0.0	0.5	0.0
Todaropsis eblanae	0.6	2.4	0.9	0.1
Unidentified cephalopod	2.3	1.8	5.6	0.9
Unidentified mollusc	0.4	0.1	0.5	0.0
Unidentified octopus	0.8	1.8	1.9	0.2
Phylum Echinodermata	**0.2**	**0.1**	**0.5**	**0.0**
Holothuroidea	0.2	0.1	0.5	0.0
Phylum Chordata: Tunicata	**0.2**	**1.2**	**0.5**	**0.0**
Unidentified tunicate	0.2	1.2	0.5	0.0
Phylum Chordata: Chondrichthyes	**1.4**	**8.3**	**3.3**	**0.2**
Rajidae	0.2	0.3	0.5	0.0
Selachii	0.2	0.4	0.5	0.0
Squalus acutipinnis	0.2	0.1	0.5	0.0
Torpedo sinuspersici	0.2	5.3	0.5	0.1
Torpedo sp.	0.2	0.8	0.5	0.0
Unidentified chondrichthyian	0.2	0.1	0.5	0.0
Unidentified elasmobranch	0.2	1.3	0.5	0.0
Phylum Chordata: Osteichthyes	**46.7**	**55.8**	**89.3**	**59.5**
Austroglossus pectoralis	0.2	0.0	0.5	0.0
Belonidae	0.2	0.4	0.5	0.0
Bregmaceros sp.	0.4	0.0	0.9	0.0
Canthigaster sp.	0.2	0.0	0.5	0.0
Chelidonichthys queketti	0.4	0.3	0.9	0.0
Chilodactylus sp.	0.2	0.7	0.5	0.0
Clupeidae	1.4	2.4	2.3	0.3
Cynoglossus zanzibarensis	0.2	0.1	0.5	0.0
Engraulis encrasicolus	16.5	7.0	23.8	20.2
Etrumeus whiteheadi	2.3	2.2	4.2	0.4
Gnathopis sp.	0.6	0.5	1.4	0.1
Gobius agulhensis	0.2	0.1	0.5	0.0
Helicolenus dactylopterus	0.4	3.5	0.9	0.1
Merluccius sp.	1.2	7.3	2.8	0.9
Paracallionymus costatus	2.1	0.9	4.2	0.5
Pterogymnus laniarius	0.2	0.2	0.5	0.0
Sardinops sagax	3.1	5.7	7.0	2.3
Scomber japonicus	0.2	3.1	0.5	0.1
Trachurus capensis	2.3	8.0	5.1	2.0
Unidentified teleost	14.4	13.4	31.8	32.7
Unidentified semi-digested material	**3.7**	**3.0**	**7.5**	**1.9**
Unidentified specimen	3.7	3.0	7.5	1.9
SUM	**100.0**	**100.0**	**176.2**	**100.0**

Appendix 14: The full stomach content dataset for large *Squalus acutipinnis* sharks sampled at the 150 m depth bin along the South coast of South Africa. A total of 228 stomachs were analysed

Species	% N	% W	% FO	% IRI
Phylum Annelida	**15.0**	**5.7**	**28.9**	**21.5**
Nereididae	14.6	5.5	27.6	21.5
Unidentified polychaete	0.4	0.2	1.3	0.0
Phylum Cnidaria	**0.4**	**1.6**	**1.3**	**0.1**
Pennatulaceae	0.4	1.6	1.3	0.1
Phylum Arthropoda	**23.1**	**4.8**	**46.5**	**6.4**
Brachyura	0.3	0.1	0.9	0.0
Caridea	1.1	0.1	3.1	0.2
Decapoda	1.4	0.2	2.2	0.1
Euphausiacea	0.4	0.0	1.3	0.0
Funchalia woodwardi	0.7	0.3	2.2	0.1
Goneplax angulata	0.4	0.3	1.3	0.0
Malacostraca	0.7	0.2	2.2	0.1
Mursia cristiata	1.0	0.4	3.1	0.2
Mysida	1.3	0.2	2.6	0.1
Parapaguridae	0.7	0.1	2.2	0.1
Pasiphaea spp. 1	0.1	0.1	0.4	0.0
Penaeidae	0.3	0.0	0.9	0.0
Plesionika martia	0.1	0.1	0.4	0.0
Pterygosquilla armata capensis	1.3	1.0	3.9	0.3
Unidentified amphipod	10.1	0.3	11.0	4.4
Unidentified crustacean	2.5	0.3	6.6	0.7
Upogebia africana	0.3	0.2	0.9	0.0
Upogebia capensis	0.3	0.2	0.9	0.0
Upogebia sp.	0.1	0.7	0.4	0.0
Phylum Mollusca	**9.4**	**15.0**	**25.4**	**3.9**
Crossia sp.	0.1	0.1	0.4	0.0
Enteroctopus magnificus	0.1	0.3	0.4	0.0
Loligo reynaudii	2.8	6.1	4.8	1.7
Inioteuthis capensis	0.1	0.0	0.4	0.0
Octopus vulgaris	0.1	0.8	0.4	0.0
Sepia australis	0.3	0.6	0.9	0.0
Sepia sp.	0.7	0.3	2.2	0.1
Sepiolidae	0.4	0.2	1.3	0.0
Teuthida	1.0	1.4	3.1	0.3
Todaropsis eblanae	0.1	0.6	0.4	0.0
Unidentified cephalopod	2.5	2.0	7.9	1.4
Unidentified mollusc	0.1	0.0	0.4	0.0
Unidentified octopus	0.8	2.6	2.6	0.3
Phylum Echinodermata	**0.3**	**0.2**	**0.9**	**0.0**
Holothuroidea	0.3	0.2	0.9	0.0
Phylum Chordata: Chondrichthyes	**0.6**	**3.9**	**1.8**	**0.3**
Unidentified elasmobranch	0.6	3.9	1.8	0.3
Phylum Chordata: Osteichthyes	**47.2**	**63.9**	**93.9**	**64.3**
Anguilliformes	0.1	0.3	0.4	0.0
Austroglossus pectoralis	0.3	0.3	0.9	0.0
Champsodontidae	0.1	0.1	0.4	0.0
Clupeidae	0.3	0.4	0.9	0.0
Cynoglossus zanzibarensis	0.3	0.2	0.9	0.0
Engraulis encrasicolus	4.8	3.6	11.4	3.7
Etrumeus whiteheadi	4.1	7.0	6.6	1.3
Gnathophis sp.	0.3	0.5	0.9	0.0
Helicolenus dactylopterus	0.3	1.6	0.9	0.1
Lampricoides	1.8	0.1	0.4	0.0
Lophius vomerinus	0.1	1.9	0.4	0.0
Merluccius capensis	0.3	0.8	0.9	0.0
Merluccius sp.	0.8	7.3	2.6	0.8
Paracallionymus costatus	13.3	3.6	12.3	8.1
Sardinops sagax	3.9	10.1	7.5	4.1
Scomber japonicus	0.3	0.8	0.9	0.0
Scomberesox saurus scomberoides	0.3	1.3	0.9	0.0
Trachurus capensis	1.8	7.9	5.3	2.0
Trichiurus lepturus	0.1	0.9	0.4	0.0
Unidentified teleost	13.9	15.2	39.0	44.0
Unidentified semi-digested material	**4.1**	**4.9**	**10.1**	**3.5**
Unidentified specimen	4.1	4.9	10.1	3.5
SUM	**100.0**	**100.0**	**208.8**	**100.0**

Appendix 15: The full stomach content dataset for large *Squalus acutipinnis* sharks sampled at the 250 m depth bin along the South coast of South Africa. A total of 7 stomachs were analysed

Species	%N	%W	% FO	% IRI
Phylum Annelida	**44.4**	**11.7**	**57.1**	**56.8**
Unidentified polychaete	44.4	11.7	57.1	56.8
Phylum Arthropoda	**22.2**	**33.1**	**28.6**	**14.0**
Parapaguridae	11.1	13.6	14.3	6.3
Unidentified crustacean	11.1	19.5	14.3	7.7
Phylum Chordata: Osteichthyes	**33.3**	**55.3**	**42.9**	**29.2**
Etrumeus whiteheadi	11.1	50.6	14.3	15.6
Unidentified teleost	22.2	4.7	28.6	13.6
SUM	**100.0**	**100.0**	**128.6**	**100.0**

Appendix 16: The full stomach content dataset for medium *Squalus bassi* sharks sampled at the 250 m depth bin along the West coast of South Africa. A total of 6 stomachs were analysed

Species	%N	%W	% FO	% IRI
Phylum Mollusca	**42.9**	**81.5**	**50.0**	**69.5**
Unidentified cephalopod	42.9	81.5	50.0	69.5
Phylum Chordata: Osteichthyes	**57.1**	**18.5**	**66.7**	**30.5**
Pegusa nasuta	14.3	17.4	16.7	5.9
Unidentified teleost	42.9	1.2	50.0	24.6
SUM	**100.0**	**100.0**	**116.7**	**100.0**

Appendix 17: The full stomach content dataset for medium *Squalus bassi* sharks sampled at the 150 m depth bin along the South coast of South Africa. A total of 24 stomachs were analysed

Species	%N	%W	% FO	% IRI
Phylum Annelida	**23.3**	**12.6**	**25.0**	**14.3**
Quill worm	2.3	1.3	4.2	0.3
Unidentified polychaete	20.9	11.3	20.8	13.9
Phylum Arthropoda	**23.3**	**10.0**	**25.0**	**5.1**
Funchalia woodwardi	7.0	0.3	4.2	0.6
Parapagurus pilosimanus	2.3	6.4	4.2	0.8
Pterygosquilla armata capensis	2.3	1.9	4.2	0.4
Unidentified crustacean	11.6	1.3	12.5	3.4
Phylum Mollusca	**20.9**	**17.2**	**37.5**	**29.7**
Unidentified cephalopod	20.9	17.2	37.5	29.7
Phylum Chordata: Osteichthyes	**23.3**	**31.0**	**33.3**	**37.5**
Unidentified teleost	23.3	31.0	33.3	37.5
Unidentified semi-digested material	**9.3**	**29.2**	**16.7**	**13.3**
Unidentified specimen	9.3	29.2	16.7	13.3
SUM		**100.0**	**100.0**	**137.5**	**100.0**

Appendix 18: The full stomach content dataset for medium *Squalus bassi* sharks sampled at the 250 m depth bin along the South coast of South Africa. A total of 6 stomachs were analysed

Species	%N	%W	% FO	% IRI
Phylum Arthropoda	**25.0**	**0.9**	**33.3**	**7.0**
Penaeidae	12.5	0.8	16.7	3.6
Pterygosquilla armata capensis	12.5	0.0	16.7	3.4
Phylum Mollusca	**12.5**	**3.2**	**16.7**	**4.3**
Todaropsis eblanae	12.5	3.2	16.7	4.3
Phylum Chordata: Osteichthyes	**62.5**	**95.9**	**83.3**	**88.7**
Merluccius paradoxus	12.5	89.7	16.7	27.7
Unidentified teleost	50.0	6.2	66.7	61.0
SUM	**100.0**	**100.0**	**133.3**	**100.0**

Appendix 19: The full stomach content dataset for medium *Squalus bassi* sharks sampled at the 450 m depth bin along the South coast of South Africa. A total of 6 stomachs were analysed

Species	%N	%W	% FO	% IRI
Phylum Arthropoda	**27.3**	**0.6**	**33.3**	**7.7**
Unidentified amphipod	18.2	0.5	16.7	5.2
Unidentified crustacean	9.1	0.0	16.7	2.5
Phylum Mollusca	**54.5**	**48.1**	**100.0**	**73.2**
Sepia australis	9.1	28.5	16.7	10.4
Sepiida	9.1	1.7	16.7	3.0
Unidentified cephalopod	36.4	18.0	66.7	59.9
Phylum Chordata: Osteichthyes	**18.2**	**51.3**	**33.3**	**19.1**
Paracallionymus costatus	9.1	4.0	16.7	3.6
Unidentified teleost	9.1	47.3	16.7	15.5
SUM	**100.0**	**100.0**	**166.7**	**100.0**

Appendix 20: The full stomach content dataset for large *Squalus bassi* sharks sampled at the 150 m depth bin along the West coast of South Africa. A total of 14 stomachs were analysed

Species	%N	%W	% FO	% IRI
Phylum Arthropoda	**13.3**	**2.3**	**14.3**	**6.1**
Pterygosquilla armata capensis	13.3	2.3	14.3	6.1
Phylum Mollusca	**13.3**	**23.3**	**28.6**	**11.2**
Loligo reynaudii	3.3	22.8	7.1	5.0
Unidentified cephalopod	10.0	0.6	21.4	6.1
Phylum Chordata: Osteichthyes	**73.3**	**74.3**	**100.0**	**82.8**
Callionymidae	23.3	2.0	7.1	4.9
Gnathophis sp.	10.0	3.5	14.3	5.2
Merluccius sp.	6.7	54.2	14.3	23.5
Paracallionymus costatus	3.3	2.6	7.1	1.1
Symbolophorus boops	6.7	1.0	7.1	1.5
Unidentified teleost	23.3	11.0	50.0	46.5
SUM	**100.0**	**100.0**	**142.9**	**100.0**

Appendix 21: The full stomach content dataset for large *Squalus bassi* sharks sampled at the 250 m depth bin along the West coast of South Africa. A total of 34 stomachs were analysed

Species	%N	%W	% FO	%IRI
Phylum Sipuncula	**2.9**	**1.0**	**5.9**	**0.6**
Sipuncula	2.9	1.0	5.9	0.6
Phylum Arthropoda	**5.9**	**0.2**	**8.8**	**0.5**
Pterygosquilla armata capensis	1.5	0.0	2.9	0.1
Malacostraca	2.9	0.0	2.9	0.2
Unidentified crustacean	1.5	0.1	2.9	0.1
Phylum Mollusca	**7.4**	**2.4**	**14.7**	**1.4**
Teuthida	1.5	1.5	2.9	0.2
Todarodes angolensis	1.5	0.7	2.9	0.2
Unidentified cephalopod	4.4	0.2	8.8	1.1
Phylum Chordata: Osteichthyes	**82.4**	**95.1**	**126.5**	**97.3**
Callionymidae	11.8	2.1	5.9	2.1
Etrumeus whiteheadi	2.9	8.6	5.9	1.8
Gnathophis sp.	2.9	2.7	5.9	0.9
Merluccius sp.	14.7	57.9	26.5	50.0
Myctophidae	5.9	0.4	8.8	1.4
Scomberesox saurus scomberoides	1.5	0.6	2.9	0.2
Symbolopholus boops	13.2	1.6	11.8	2.0
Trachurus capensis	4.4	7.7	8.8	2.8
Trichiurus lepturus	2.9	5.2	5.9	1.2
Unidentified teleost	22.1	8.4	44.1	34.9
Unidentified semi-digested material	**1.5**	**1.3**	**2.9**	**0.2**
Unidentified invertebrate	1.5	1.3	2.9	0.2
SUM	**100.0**	**100.0**	**158.8**	**100.0**

Appendix 22: The full stomach content dataset for large *Squalus bassi* sharks sampled at the 350 m depth bin along the West coast of South Africa. A total of 13 stomachs were analysed

Species	%N	%W	% FO	% IRI
Phylum Arthropoda	**8.0**	**0.2**	**15.4**	**1.3**
Pterygosquilla armata capensis	4.0	0.2	7.7	0.7
Decapoda	4.0	0.0	7.7	0.6
Phylum Mollusca	**20.0**	**22.4**	**38.5**	**11.2**
Todaropsis eblanae	4.0	11.5	7.7	2.4
Todarodes sp.	4.0	7.7	7.7	1.8
Unidentified cephalopod	12.0	3.1	23.1	7.0
Phylum Chordata: Osteichthyes	**68.0**	**75.6**	**92.3**	**86.6**
Etrumeus whiteheadi	12.0	1.5	7.7	2.1
Merluccius sp.	16.0	54.6	15.4	21.8
Symbolophorus sp.	8.0	2.0	7.7	1.5
Unidentified teleost	32.0	17.6	61.5	61.2
Phylum Chordata: Mammalia	**4.0**	**1.8**	**7.7**	**0.9**
Unidentified mammal	4.0	1.8	7.7	0.9
SUM	**100.0**	**100.0**	**153.8**	**100.0**

Appendix 23: The full stomach content dataset for large *Squalus bassi* sharks sampled at the 550 m depth bin along the West coast of South Africa. A total of 12 stomachs were analysed

Species	%N	%W	% FO	%IRI
Phylum Mollusca	**19.2**	**6.4**	**41.7**	**8.8**
Teuthida	7.7	5.0	16.7	3.5
Unidentified cephalopod	11.5	1.4	25.0	5.3
Phylum Chordata: Osteichthyes	**80.8**	**93.6**	**141.7**	**91.2**
Lophiiformes	7.7	1.6	16.7	2.5
Merluccius sp.	7.7	23.5	16.7	8.5
Sardinops sagax	3.8	1.8	8.3	0.8
Symbolophorus sp.	3.8	0.6	8.3	0.6
Trichiurus lepturus	19.2	21.8	33.3	22.4
Thyrsites atun	3.8	13.0	8.3	2.3
Unidentified teleost	34.6	31.3	50.0	54.1
SUM	**100.0**	**100.0**	**183.3**	**100.0**

Appendix 24: The full stomach content dataset for large *Squalus bassi* sharks sampled at the 150 m depth bin along the South coast of South Africa. A total of 57 stomachs were analysed

Species	%N	%W	% FO	%IRI
Phylum Arthropoda	**4.0**	**0.2**	**10.5**	**0.7**
Decapoda	0.7	0.0	1.8	0.1
Malacostraca	1.3	0.1	3.5	0.3
Pterygosquilla armata capensis	0.7	0.0	1.8	0.1
Unidentified crustacean	1.3	0.1	3.5	0.3
Phylum Mollusca	**51.7**	**28.9**	**40.4**	**21.4**
Crossia sp.	0.7	0.1	1.8	0.1
Histioteuthis macrohista	0.7	0.2	1.8	0.1
Histioteuthis sp.	0.7	0.0	1.8	0.1
Loligo reynaudii	6.6	9.7	5.3	4.7
Sepia australis	32.5	7.4	3.5	7.7
Sepia sp.	1.3	2.3	3.5	0.7
Teuthida	1.3	1.2	3.5	0.5
Todarodes angolensis	0.7	6.0	1.8	0.6
Unidentified cephalopod	6.0	1.7	15.8	6.7
Unidentified octopus	1.3	0.2	1.8	0.1
Phylum Chordata: Osteichthyes	**43.0**	**69.5**	**82.5**	**77.7**
Coelorhinchus sp.	0.7	7.8	1.8	0.8
Engraulis encrasicolus	0.7	0.1	1.8	0.1
Etrumeus whiteheadi	2.0	1.1	5.3	0.5
Gnathophis sp.	2.0	4.8	5.3	2.0
Lepidopus sp.	1.3	8.7	3.5	1.9
Merluccius paradoxus	0.7	3.7	1.8	0.4
Merluccius sp.	1.3	3.3	3.5	0.9
Myctophidae	4.0	1.2	1.8	0.5
Paracallionymus costatus	3.3	0.3	1.8	0.3
Sardinops sagax	3.3	4.4	5.3	2.3
Satyrichthys adeni	0.7	0.1	1.8	0.1
Scomberesox saurus scomberoides	2.0	2.1	5.3	0.7
Tharbacus sp.	0.7	0.6	1.8	0.1
Trichiurus lepturus	2.6	1.8	7.0	1.7
Trachurus capensis	3.3	6.0	5.3	2.7
Unidentified teleost	14.6	23.5	29.8	62.6
Unidentified semi-digested material	**1.3**	**1.4**	**3.5**	**0.3**
Unidentified invertebrate	0.7	0.2	1.8	0.1
Unidentified specimen	0.7	1.2	1.8	0.2
SUM	**100.0**	**100.0**	**136.8**	**100.0**

Appendix 25: The full stomach content dataset for large *Squalus bassi* sharks sampled at the 250 m depth bin along the South coast of South Africa. A total of 31 stomachs were analysed

Species	%N	%W	% FO	% IRI
Phylum Annelida	**2.3**	**0.3**	**3.2**	**0.1**
Pseudonereis variegata	2.3	0.3	3.2	0.1
Phylum Arthropoda	**9.3**	**0.8**	**12.9**	**1.2**
Penaeidae	7.0	0.8	9.7	1.1
Pterygosquilla armata capensis	2.3	0.0	3.2	0.1
Phylum Mollusca	**9.3**	**4.4**	**12.9**	**2.6**
Unidentified cephalopod	9.3	4.4	12.9	2.6
Phylum Chordata: Osteichthyes	**76.7**	**91.1**	**96.8**	**95.8**
Etrumeus whiteheadi	2.3	5.8	3.2	0.4
Gnathophis sp.	2.3	0.3	3.2	0.1
Lepidopus sp.	2.3	18.5	3.2	1.0
Myctophidae	4.7	6.7	6.5	1.1
Sardinops sagax	2.3	0.5	3.2	0.1
Scomber japonicus	4.7	12.0	6.5	1.6
Symbolophorus boops	2.3	4.2	3.2	0.3
Trichiurus lepturus	2.3	1.0	3.2	0.2
Unidentified teleost	53.5	42.0	64.5	91.0
Unidentified semi-digested material	**2.3**	**3.5**	**3.2**	**0.3**
Unidentified specimen	2.3	3.5	3.2	0.3
SUM	**100.0**	**100.0**	**129.0**	**100.0**

Appendix 26: The full stomach content dataset for large *Squalus bassi* sharks sampled at the 450 m depth bin along the South coast of South Africa. A total of 87 stomachs were analysed

Species	%N	%W	% FO	% IRI
Phylum Porifera	**0.6**	**0.0**	**1.1**	**0.0**
Unidentified sponge	0.6	0.0	1.1	0.0
Phylum Arthropoda	**10.8**	**0.9**	**17.2**	**2.4**
Caridea	0.6	0.0	1.1	0.0
Funchalia woodwardi	6.6	0.7	9.2	2.2
Mursia cristiata	0.6	0.0	1.1	0.0
Penaeidae	0.6	0.0	1.1	0.0
Unidentified amphipod	1.2	0.0	2.3	0.1
Unidentified crustacean	1.2	0.1	2.3	0.1
Phylum Mollusca	**39.2**	**14.6**	**63.2**	**41.3**
Loligo reynaudii	3.0	3.4	4.6	0.5
Lycoteuthis lorigera	1.8	1.2	3.4	0.3
Sepia australis	6.0	1.1	2.3	0.5
Sepia hieronis	0.6	0.4	1.1	0.0
Sepia sp.	0.6	0.1	1.1	0.0
Teuthida	1.2	0.1	2.3	0.1
Todaropsis eblanae	0.6	0.1	1.1	0.0
Unidentified cephalopod	23.5	4.4	43.7	39.4
Unidentified octopus	1.8	3.7	3.4	0.3
Phylum Chordata: Osteichthyes	**48.8**	**84.1**	**80.5**	**56.2**
Diaphus sp.	1.2	0.2	2.3	0.1
Engraulis encrasicolus	1.2	0.4	2.3	0.1
Epigonus sp.	0.6	0.5	1.1	0.0
Etrumeus whiteheadi	3.0	2.4	4.6	0.6
Helicolenus dactylopterus	0.6	0.9	1.1	0.1
Lampricoides	0.6	0.0	1.1	0.0
Merluccius paradoxus	3.6	12.9	6.9	3.7
Merluccius sp.	1.8	4.1	3.4	0.7
Myctophidae	3.6	0.6	5.7	0.5
Paracallionymus costatus	1.8	0.1	3.4	0.2
Sardinops sagax	1.2	0.6	2.3	0.1
Scomber japonicus	4.8	18.7	4.6	1.7
Scomberesox saurus scomberoides	1.2	0.2	2.3	0.1
Thyrsites atun	0.6	0.7	1.1	0.0
Trichiurus lepturus	1.2	9.0	2.3	0.8
Trachurus capensis	2.4	3.2	3.4	0.6
Unidentified teleost	18.7	28.0	31.0	46.8
Zeus capensis	0.6	1.5	1.1	0.1
Unidentified semi-digested material	**0.6**	**0.4**	**1.1**	**0.0**
Unidentified specimen	0.6	0.4	1.1	0.0
SUM	**100.0**	**100.0**	**163.2**	**100.0**

Appendix 27: The full stomach content dataset for large *Squalus bassi* sharks sampled at the 550 m depth bin along the South coast of South Africa. A total of 55 stomachs were analysed

Species	%N	%W	% FO	% IRI
Phylum Arthropoda	**12.6**	**1.9**	**20.0**	**2.1**
Aristaeomorpha foliaceae	2.9	0.2	5.5	0.6
Funchalia woodwardi	3.9	0.3	5.5	0.8
Merhippolyte agulhasensis	1.0	0.1	1.8	0.1
Pasiphaea spp. 1	1.0	0.2	1.8	0.1
Unidentified crustacean	2.9	1.2	3.6	0.5
Unidentified ostracod	1.0	0.0	1.8	0.1
Phylum Mollusca	**32.0**	**15.1**	**50.9**	**38.7**
Loligo reynaudii	3.9	2.1	5.5	1.2
Sepia australis	1.0	1.1	1.8	0.1
Todarodes angolensis	3.9	3.8	5.5	1.5
Todaropsis eblanae	2.9	0.8	1.8	0.2
Unidentified cephalopod	20.4	7.2	36.4	35.7
Phylum Chordata: Osteichthyes	**48.5**	**71.1**	**76.4**	**50.7**
Coelorhinchus sp.	1.0	1.3	1.8	0.1
Engraulis encrasicolus	2.9	3.0	3.6	0.8
Etrumeus whiteheadi	2.9	3.8	3.6	0.4
Gnathophis sp.	1.9	0.3	3.6	0.3
Helicolenus dactylopterus	1.0	1.2	1.8	0.1
Lampricoides	1.0	0.1	1.8	0.1
Merluccius sp.	6.8	32.0	12.7	17.5
Muraenidae	1.0	3.2	1.8	0.3
Mullidae	1.9	4.9	1.8	0.4
Myctophidae	1.9	0.9	3.6	0.4
Paracallionymus costatus	1.9	0.1	3.6	0.3
Sardinops sagax	1.0	1.6	1.8	0.2
Symbolophorus boops	1.0	0.1	1.8	0.1
Trichiurus lepturus	1.9	7.8	3.6	1.3
Unidentified teleost	19.4	9.8	27.3	28.3
Zeus capensis	1.0	1.0	1.8	0.1
Unidentified semi-digested material	**6.8**	**11.9**	**12.7**	**8.5**
Unidentified specimen	6.8	11.9	12.7	8.5
SUM	**100.0**	**100.0**	**160.0**	**100.0**

Appendix 28: The full stomach content dataset for large *Squalus bassi* sharks sampled at 650 m depth bin along the South coast of South Africa. A total of 6 stomachs were analysed

Species	%N	%W	% FO	% IRI
Phylum Mollusca	**70.0**	**83.8**	**100.0**	**86.9**
Todarodes angolensis	20.0	58.4	33.3	44.3
Todaropsis eblanae	20.0	12.3	33.3	18.2
Unidentified cephalopod	30.0	13.2	33.3	24.4
Phylum Chordata: Osteichthyes	**30.0**	**16.2**	**33.3**	**13.1**
Gnathophis sp.	20.0	11.5	16.7	8.9
Unidentified teleost	10.0	4.7	16.7	4.2
SUM	**100.0**	**100.0**	**133.3**	**100.0**

Appendix 29: The sum (N) and mean isotopic expression of Nitrogen 15 (δ^{15}N‰) and Carbon 13 (δ^{13}C‰; expressed by delta notation, in parts per thousand) of *Squalus acutipinnis* and *S. bassi* sampled on the South and West coast of South Africa. Means are categorised by coast, depth (in meters) and size classes

		South coast		West coast	
		S. acutipinnis	*S. bassi*	*S. acutipinnis*	*S. bassi*
By coast:	N	68	75	25	28
	δ^{15}N (‰)	14.22	14.04	14.83	14.88
	±SE	0.081	0.108	0.129	0.114
	δ^{13}C (‰)	-15.10	-15.75	-15.85	-16.04
	±SE	0.065	0.087	0.075	0.054
		≤ 200 m			
	N	55	20	8	6
	δ^{15}N (‰)	14.05	14.15	14.51	14.76
	±SE	0.082	0.213	0.228	0.369
	δ^{13}C (‰)	-14.97	-15.01	-16.06	-15.98
	±SE	0.068	0.148	0.135	0.113
By depth:		> 200 m			
	N	13	55	17	22
	δ^{15}N (‰)	14.95	14.18	14.98	14.92
	±SE	0.093	0.127	0.147	0.110
	δ^{13}C (‰)	-15.65	-16.15	-15.75	-16.06
	±SE	0.070	0.071	0.082	0.063
		small			
	N	22	11	2	3
	δ^{15}N (‰)	13.41	13.26	14.63	13.94
	±SE	0.132	0.148	0.271	0.310
	δ^{13}C (‰)	-14.78	-14.70	-15.88	-15.97
	±SE	0.098	0.238	0.091	0.132
By size:		large			
	N	46	64	23	25
	δ^{15}N (‰)	14.47	14.17	14.85	15
	±SE	0.080	0.117	0.138	0.103
	δ^{13}C (‰)	-15.25	-16.04	-15.85	-16.05
	±SE	0.074	0.069	0.081	0.058